Biomarkers and Human Biomonitoring

Volume 2: Selected Biomarkers of Current Interest

Issues in Toxicology

Series Editors:
Professor Diana Anderson, *University of Bradford, UK*
Dr Michael D Waters, *Integrated Laboratory Systems, Inc, N Carolina, USA*
Dr Martin F Wilks, *University of Basel, Switzerland*
Dr Timothy C Marrs, *Edentox Associates, Kent, UK*

Titles in the Series:
1: Hair in Toxicology: An Important Bio-Monitor
2: Male-mediated Developmental Toxicity
3: Cytochrome P450: Role in the Metabolism and Toxicity of Drugs and other Xenobiotics
4: Bile Acids: Toxicology and Bioactivity
5: The Comet Assay in Toxicology
6: Silver in Healthcare
7: *In Silico* Toxicology: Principles and Applications
8: Environmental Cardiology
9: Biomarkers and Human Biomonitoring, Volume 1: Ongoing Programs and Exposures
10: Biomarkers and Human Biomonitoring, Volume 2: Selected Biomarkers of Current Interest

How to obtain future titles on publication:
A standing order plan is available for this series. A standing order will bring delivery of each new volume immediately on publication.

For further information please contact:
Book Sales Department, Royal Society of Chemistry, Thomas Graham House, Science Park, Milton Road, Cambridge, CB4 0WF, UK
Telephone: +44 (0)1223 420066, Fax: +44 (0)1223 420247, Email: books@rsc.org
Visit our website at http://www.rsc.org/Shop/Books/

Biomarkers and Human Biomonitoring

Volume 2: Selected Biomarkers of Current Interest

Edited by

Lisbeth E. Knudsen
Institute of Public Health, University of Denmark, Denmark

Domenico Franco Merlo
Epidemiology and Biostatistics, Istituto Nazionale per la Ricerca sul Cancro, Genova, Italy

RSC Publishing

Issues in Toxicology No. 10

ISBN: 978-1-84973-242-0
ISSN: 1757-7179

A catalogue record for this book is available from the British Library

Published by The Royal Society of Chemistry,
Thomas Graham House, Science Park, Milton Road,
Cambridge CB4 0WF, UK

Registered Charity Number 207890

For further information see our web site at www.rsc.org

Printed in Great Britain by CPI Group (UK) Ltd, Croydon, CR0 4YY

Preface: Handbook of Biomarkers in Environmental Health

Human biomonitoring has developed from a research tool in occupational and environmental health to identify and quantify exposures to harmful substances in urine or blood. The analytical methods for detection of substances in biological media have been considerably improved with smaller detection limits and more precise and specific measurements.

Human biomonitoring is a valuable tool for exposure estimation in selected populations and is currently used in surveillance programs all over the world. In autumn 2011 a European harmonization of human biomonitoring in schoolchildren and their mothers will be initiated in 17 countries. The measurements include mercury in hair and cadmium, cotinine, and phthalates in urine in pairs of 120 children and their mothers in each country. Some countries also include bisphenol A, triclosan, and parabens in urine.

This handbook describes the current human biomonitoring activities in Germany, Romania, France, Canada, India, and Belgium, providing convincing evidence of a global decline in human exposures to lead and increasing concern related to adverse effects from exposures to endocrine disruptors and genotoxic compounds. The book includes chapters on measurement of human exposure to phthalates, perfluorinated compounds (PFCs), bisphenol A, brominated flame retardants, lead, polycyclic aromatic hydrocarbons (PAHs), dioxins, mercury, and arsenic using biomarkers of exposure. Chapters describing human biomonitoring of exposures to environmental tobacco smoke, mycotoxins, physiological stress, hormone activity, oxidative stress, and ionizing radiation are included, as well as chapters on effect biomarkers of hemoglobin adducts, germ cells, micronuclei, and of individual susceptibility. A further chapter describes the ethical issues related to human sampling and monitoring.

Issues in Toxicology No. 10
Biomarkers and Human Biomonitoring, Volume 2: Selected Biomarkers of Current Interest
Edited by Lisbeth E. Knudsen and Domenico Franco Merlo

Published by the Royal Society of Chemistry, www.rsc.org

In Germany, two main human biomonitoring studies have been ongoing since the mid-1980s. The German Environmental Survey (GerES) has been carried out four times, with adults, with children or with both adults and children. The second, the Environmental Specimen Bank (ESB), has routinely collected human specimens from students every year which today provide a historical record of exposures.

In Romania, specific exposures and specific populations have been studied by human biomonitoring. A study on lead in young children was conducted over a period of 10 years, and from this an intervention program was implemented to reduce exposure to lead. In a second study on lead, both pregnant women and young children were examined for exposure to lead, while studies on arsenic in drinking water were carried out in a geographically bounded population.

In France, human biomonitoring studies have until recently focused on specific exposures or populations. Now, a multipollutant study on a national level involving all age groups has begun.

There have been several human biomonitoring studies in Canada, but four major studies are currently running which examine different exposures and populations: the nationally representative Canadian Health Measures Survey (CHMS), the Canada-wide Maternal–Infant Research on Environmental Chemicals (MIREC) study, the research and community-based Northern Contaminants Program (NCP), and mercury biomonitoring of First Nations people living on reserves south of the 60th parallel.

The Indian population is of great diversity, leading to differences in exposure to pollutants. Studies have been carried out in different populations and many different exposures, such as mercury, lead, and arsenic, have been examined. Furthermore, there have been studies on biomarkers of effect and susceptibility.

In Belgium, a large human biomonitoring study, the second Flemish human biomonitoring survey, has run in the period 2007–2011. This study included a representative sample of the country's population with more than 200 people from each of the following groups: newborns and their mothers, adolescents, and adults. Cord blood, blood, urine, and hair samples were collected and analyzed for a wide range of pollutants such as heavy metals, persistent chlorinated compounds, 1-hydroxy pyrene, and *t,t*-muconic acid, brominated flame retardants, metabolites of bisphenol A, parabens, organophosphate pesticide para-hydroxybenzoic acid, and para-dichlorobenzene.

On a European Union level, COPHES are cross-national programs with 27 countries participating in work towards a functional framework and standards for coherent human biomonitoring (HBM). Furthermore, COPHES will be a step towards improvement of exposure quantification and towards establishing HBM as a tool for the environmental and health policy.

Human biomonitoring measurements on phthalates in urine are an ideal tool because this determines internal exposures by measuring metabolites, and thus is representative for exposure from all sources and by all routes. This has opened a new and alternative approach to risk assessment for phthalates.

Perfluorinated compounds (PFCs) are widely used chemicals, and are found both in industry and in private homes. Possible routes of exposure are eating or drinking contaminated foods, inhalation, or direct contact with PFC-coated products. Human biomonitoring contributes to the integrated measurement of exposure to PFCs and accumulated PFCs, because these chemicals are persistent in humans. Measurement of levels in serum has been recognized as the most suitable biomarker for PFCs. Some PFCs have been phased out or controlled for emissions since around the year 2000, leading to declines in serum concentrations.

Dioxins are unwanted by-products formed in a variety of industrial and thermal processes. Many biomarkers of dioxins are not specific, such as liver enzymes, blood lipids, thyroid and steroid hormones, chloracne, glucose tolerance, and cytochrome P450 (CYP) induction. In combination with chemical analysis of dioxins in blood or adipose tissue, nonspecific biomarkers of dioxin exposure can be useful, but they should not be used as an isolated tool for risk assessment.

Brominated flame retardants (BFRs) are known to act as an endocrine disrupter. They have been extensively used, but they are now partly regulated by law or voluntarily withdrawn from the market due to persistent and accumulative properties. The most important exposure pathways of BFRs have been found to be diet and the indoor environment. In human biomonitoring studies, blood and breast milk are commonly used samples to measure metabolites of BFRs, but hair is also a promising specimen.

The emission of lead declined significantly in European countries in the period 1990–2004, but in spite of this, lead can still be monitored in humans. Blood is the most widely used matrix to measure exposure of lead. Exposure to lead is known to cause adverse effects in the central nervous system, especially in children where an increase of lead in blood is associated with a decrease in IQ.

Polycyclic aromatic hydrocarbons (PAHs) are known to have carcinogenic, reprotoxic, and immunotoxic health effects. Exposure pathways are through food, dermal absorption, and inhalation. Biomarkers of PAH have been studied for many years and today the urinary biomarker 1-hydroxypyrene (HO-Pyr) has been found to be the most suitable biomarker.

Bisphenol A (BPA) is widely used in many plastic products, especially for food or beverage packaging. BPA is known to be an endocrine disrupter, and is regulated in some countries, especially in products for infants and small children. In human biomonitoring studies BPA is mostly measured in urine, but it can also be measured in blood, breast milk, follicular and amniotic fluids, placenta, and umbilical cord blood measurements revealing fetal exposure.

Human biomonitering of mercury depends on the species of interest. Elemental mercury can be measured in urine and in blood because it increases the concentration of mercury in the plasma. Methyl mercury can be measured in scalp hair and in the blood where it accumulates mainly in the red blood cells.

Hemoglobin (Hb) adducts reflect the dose of an electrophilic compound in the red blood cells. Hemoglobin adducts have been used to measure general background exposure to certain compounds (such as aromatic amines, epoxides, and isocyanates) and also to assess clinical and occupational exposures.

Human biomonitoring for environmental tobacco smoke (ETS) can be used to identify people at risk and to assess whether tobacco control strategies result in less ETS. Widely used biomarkers are nicotine and its metabolites in blood, urine, and saliva, but other biomarkers in urine are also used. Biomarkers of long-term exposure can be measured in hair and nails. In addition to these specimens, measurements of exposure in children can be carried out using meconium, breast milk, and deciduous teeth.

Mycotoxins, such as aflatoxins and fumonisins, are frequently found in food, and people with homogeneous diets are at greatest risk. Mycotoxins are known to affect the immune system, growth, and the development of cancer. Biomarkers have been used to understand the molecular epidemiology of chronic disease and may be used in the future to understand the etiology of diseases.

Investigations into the biological pathways linking stress and health are of increasing interest in society, the working environment, and individuals. Endocrine factors have become increasingly relevant for the understanding of adaptation processes and the pathogenesis of chronic diseases caused by adverse psychosocial working environments. An intricate network of hormones and hormone-like activities is implicated in the stress response, and current research programs identify and validate suitable biomarkers for physiological stress.

Micronuclei (MN) are small extranuclear bodies that arise in dividing cells from acentric chromosome/chromatid fragments or whole chromosomes/chromatids lagging behind in anaphase that are not included in the daughter nuclei at telophase. The MN assay allows detection of both aneugens and clastogens, shows simplicity of scoring, is widely applicable in different cell types, is internationally validated, has potential for automation, and is predictive for cancer. The cytokinesis-block micronucleus assay (CBMN) has been widely applied in HBM studies of populations exposed to occupational and environmental mutagens.

The link between genotypes, which take into account inter-individual differences in response to genotoxic exposure, and the occurrence of cytogenetic damage, which quantifies the extent of genetic damage due to environmental and occupational exposures, has been extensively reported in the literature. Although it is clear that MN induction can be affected by genetic polymorphisms in various genes involved in DNA repair pathways, xenobiotic metabolism, and folate metabolism, more and larger-scale studies are required to better understand MN formation driven by genetic polymorphisms, especially for aneuploidy-related genetic polymorphisms.

Humans are exposed to a wide range of endocrine disrupting chemicals, and blood is an important specimen when measuring the integrated effect of the chemicals. To measure hormone activity, *ex vivo* cell systems have been introduced.

The assessment of exposure of germ cells is used to determine genomic, transcriptional, and translational errors in the cell itself and to determine whether these alterations can be transferred to the offspring. Biomarkers include aberrations at the chromosomal as well as at the nucleotide level.

Biomarkers of oxidative stress measure oxidative damage to DNA. In lipids, tissues, or biological fluids a correlation is seen between oxidative stress and exposure to traffic generated air pollution, although these biomarkers are not specific for either exposure or disease.

Exposure to ionizing radiation can be detected in human bodies because it can produce genome and epigenetic damage, disturbances of signal molecules or cell proteins, and temporary or permanent disturbance to biological pathways. The measurement of genetic damage after radiation exposure is the most commonly used human biomarker historically, and provides a warning when limits of exposure have been exceeded.

Diverse domestic implementations of EU regulations or international guidelines may hinder transnational research and bring about inequalities in the level of protection in different countries. Some forthcoming challenges and proposed solutions, regarding communication, secondary use, and exchange of data, are presented in the chapter on ethics.

Nanna Hundebøll
Lisbeth E. Knudsen
Domenico Franco Merlo

Contents

Volume 1

Issues in Toxicology No. 10
Biomarkers and Human Biomonitoring, Volume 2: Selected Biomarkers of Current Interest
Edited by Lisbeth E. Knudsen and Domenico Franco Merlo

Published by the Royal Society of Chemistry, www.rsc.org

Volume 2

CHAPTER 4

Biomarkers of Exposure: Hemoglobin Adducts

H. VON STEDINGK,[a] S. OSTERMAN-GOLKAR[b] AND M. TÖRNQVIST*,[a]

[a] Stockholm University, Department of Materials and Environmental Chemistry, Arrhenius Laboratories of Natural Sciences, 106 91 Stockholm, Sweden; [b] Stockholm University, Department of Genetics, Microbiology and Toxicology, Arrhenius Laboratories of Natural Sciences, 106 91 Stockholm, Sweden

4.1 Introduction

Chemical compounds that are electrophilically reactive or form reactive intermediates in the metabolism are potentially toxic owing to their ability to react with electron-dense atoms in proteins and DNA. The reactivity makes these compounds difficult to measure *in vivo*, as a result of their fast detoxification and short half-lives.

One approach used for measurement of exposure or the *in vivo* dose of electrophilic compounds is based on measurement of their stable adducts with biomacromolecules. The macromolecules used for this purpose are mostly DNA, serum albumin and hemoglobin (Hb) in blood. An adduct has been defined as a complex that forms when a chemical binds to a biological molecule. In the present context it has also been found practical to use the term adduct for a moiety covalently bound to the macromolecule as a consequence of a reaction (Figure 4.1).

Issues in Toxicology No. 10
Biomarkers and Human Biomonitoring, Volume 2: Selected Biomarkers of Current Interest
Edited by Lisbeth E. Knudsen and Domenico Franco Merlo

Published by the Royal Society of Chemistry, www.rsc.org

The emphasis of this chapter is on the application of adducts with Hb as a biomarker of occupational, environmental and lifestyle exposures to carcinogenic compounds. Some applications are illustrated with examples from the authors' own work. The usefulness of adduct studies is by no means restricted to mutagens and carcinogens. The chemical reactivity of compounds and/or of their metabolic intermediates may, even at very low concentrations, alter tissue constituents in such a way as to give rise to various harmful conditions or diseases.

As discussed later in more detail the Hb adduct level reflects the dose (AUC; area under the concentration *vs.* time curve) of an electrophilic compound in blood. The AUC is determined by the absorption (or formation) of the compound, distribution, metabolism and excretion (Figure 4.2).

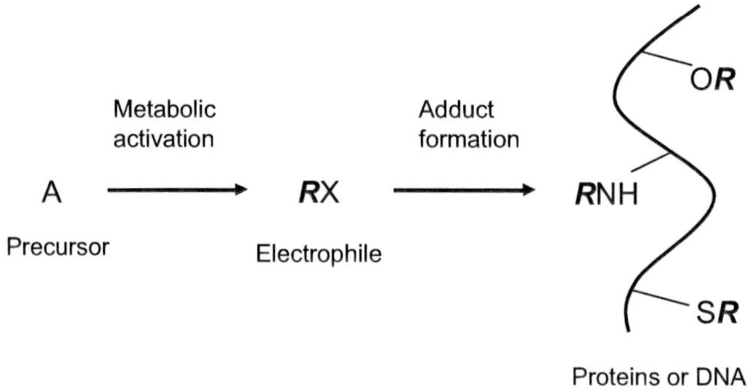

Figure 4.1 Electrophilically reactive compounds (*R*X) react with nucleophilic atoms in DNA (O, N) and proteins (O, N, S) and form adducts (*R*).

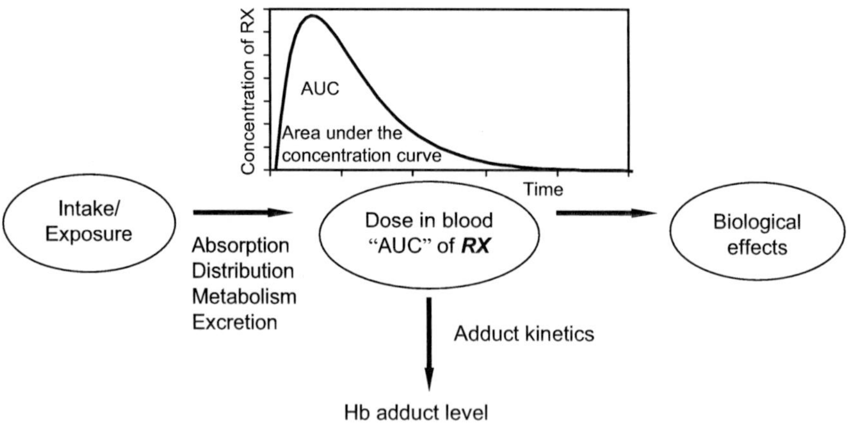

Figure 4.2 The measured Hb adduct level reflects the AUC in blood of a reactive compound. The AUC reflects all processes with an influence on the concentration over time of the reactive compound.

4.1.1 Early Studies/Milestones

By the middle of the twentieth century, adducts with proteins, including Hb, were used to measure the bioavailability of reactive compounds in experimental animals. Notably, studies with aromatic amines contributed to this field of research. In 1953, Jackson and Thompson[1] demonstrated that a radiolabeled derivative of phenylhydroxylamine bound strongly to Hb in erythrocytes and that this bound material was eliminated only in the course of erythrocyte degradation. The usefulness of monitoring Hb adducts from aromatic amines was later supported by studies of occupational exposure.[2,3]

In 1974, Ehrenberg and co-workers suggested that *in vivo* doses from exposure to electrophilic agents might be determined through the measurement of the adducts they form with tissue proteins,[4] leading to the exploration of Hb as a monitor molecule. In experiments with mice the approach was tested with the directly reactive agent ethylene oxide, and a compound that requires metabolic activation, namely dimethylnitrosamine.[5] The early development of analytical methods for the isolation of adducts, and for their analysis by gas chromatography–mass spectrometry (GC-MS), was important for the application of Hb adduct measurement to exposure to potentially genotoxic and cancer risk increasing agents. Adducts with Hb in humans were used for the first time by Calleman *et al.*[6] to calculate doses in blood explicitly following occupational exposure to ethylene oxide.

4.2 Mechanisms and Kinetics of Adduct Formation

4.2.1 Hemoglobin and Reactive Sites in Hemoglobin

Hemoglobin (Hb) is synthesized in the erythrocytes during their development in the bone marrow. The normal human adult Hb molecule, $\alpha_2\beta_2$ (HbA), is a tetramer consisting of two α-chains and two β-chains, each containing a heme group. In "normal" adults hemoglobin HbA is the dominating component, making up ~97% of the total Hb substance.

The major sites of adduct formation are cysteine-S, histidine-N, the N-terminal NH_2 group, and the carboxyl groups of aspartic and glutamic acid and of the C-terminal amino acid. Further, serine, threonine, tyrosine, lysine, arginine, methionine, and tryptophan residues may react with electrophilic compounds. Valine is the N-terminal amino acid of both the α- and the β-chain of adult Hb.

The pattern of binding of an electrophilic compound to the various nucleophilic sites in a protein may, at least to some extent, be predicted. The reactivity of nucleophilic atoms towards alkylating agents generally decreases in the order $S > N > O$. Alkylating agents have been assigned s-values[7] that describe their ability to react selectively with these atoms. Alkylating agents with high s-values, such as ethylene oxide, have a strong propensity for reactions with cysteine-S. Agents with a low s-value, such as the reactive intermediates of alkylnitrosamines, show less selectively and give relatively high levels of adducts with reactive oxygen atoms.

The reactivity of a nucleophilic site in Hb is also dependent on its pK_a; the base in an acid–base equilibrium being far more reactive than the acid (reviewed by Törnqvist *et al.*[8]). This influences the rate of adduct formation. Thus, because of the low pK_a of the N-terminal amino groups in hemoglobin (6.8–7.8), histidines (5.6–7), and cysteines (7.9–8.5) these sites have a comparatively high reactivity towards, for example, alkylating agents. Lysines have high pK_a-values (9.5–12.5) and consequently have a relatively low reactivity towards this type of electrophilic compound. However, there is also a correlation between alkalinity (as measured by pK_a) and nucleophilic strength, counteracting the effect of protonation. Thus, certain types of electrophilic compound, such as acylating agents, may give relatively high levels of adducts with lysine. A majority of the applications are based on determination of adducts with cysteine and N-terminal valine.

For high-molecular-weight compounds, it may be difficult to predict the sites of reaction. It has been suggested that the tertiary structure of the protein may be playing a more important role in adduct formation for these compounds.[9]

Neonates: At birth, HbF ($\alpha_2\gamma_2$) comprises a major part of the child's Hb. These levels decline and after 6 months adult Hb ($\alpha_2\beta_2$) takes over as the predominant form of Hb in normal children. In HbF the N-terminal amino acid of the γ-chain is glycine. Such differences in Hb should be considered in biomonitoring of Hb adducts in neonates.

Other species: There is a high degree of homology in the amino acid sequences in Hb of different mammalian species. Valine is the N-terminal amino acid in the α- and β-chain of several species. This amino acid residue has a similar reactivity in mouse, rat, dog and human Hb as has been shown in *in vitro* experiments with various low-molecular-weight epoxides and with acrylonitrile and acrylamide.[10] Thus, besides a correction for species differences in the life span of erythrocytes in the laboratory animals used, comparisons of the dose of reactive compounds could be based on measurements of adducts with N-terminal valine, with only minor correction for differences in reactivity. However, the Hb from some other species studied, for instance bovine Hb, has N-terminal valines in two of the four chains.

The Hb of the rat and some but not all mouse strains contains a cysteine residue with a particularly high reactivity. Segerbäck[11] compared the *in vitro* rates of reaction of ethylene oxide with Hb and found 170 and 12 times higher reactivity of cysteine in rat and mouse Hb, respectively, than in human Hb. Species differences in Hb binding *in vivo* that are attributed to differences in cysteine reactivity have been demonstrated in several studies, for instance for acrylonitrile in rats and mice.[12] This difference in reactivity has to be accounted for in interspecies comparisons.

4.2.2 Accumulation of Adducts through Formation and Removal

The life span of erythrocytes with their content of Hb is generally considered to be about 4 months in humans. Thus, the level of Hb adducts observed in a blood sample is the result of adduct formation and adduct removal during the 4 months prior to sampling [Scheme 4.1, Equation (4.1)].

The yield of adduct formation in Hb depends on the concentration and the persistence (AUC) of the electrophilic compound within the erythrocytes. Further, the yield depends on the chemical reactivity of the electrophile towards the nucleophilic sites involved in the adduct formation. The relationship between adduct yield and AUC for a single exposure of relatively short duration is shown in Scheme 4.1, Equation (4.2).

The basic equation for adduct accumulation

Mathematically, the adduct level measured at a certain day **n** can be expressed as the sum of contributions, from each of the preceding days **i**, of the erythrocyte life span ($t_{er} = 122$) according to:

$$A_{acc} = \sum_{(n-i)=1}^{t_{er}} (\alpha_i \times B_{n,i}) \qquad 0 < (n-i) \leq t_{er} \qquad (4.1)$$

Here, α_i is the incremental increase in adduct level formed at day **i** and $B_{n,i}$ is a factor that accounts for detraction of this increment during the time period between day **i** and day **n**.[13]

The adduct increment α_i is related to the dose (AUC) over the period of day **i** according to

$$\alpha_i = k_Y \times AUC_i \qquad (4.2)$$

where k_Y is the second-order reaction rate constant for formation of the adduct. This rate constant may be determined in *in vitro* experiments through incubation of erythrocytes or whole blood with the electrophilic compound of interest.

The loss of adducts through various processes is described with factor $B_{n,i}$ in equation (4.1):

$$B_{n,i} = (1 - [(n-i)/122]) \times f \quad 0 < (n-i) \leq t_{er} \qquad (4.3)$$

The first factor in this equation, $(1 - [(n-i)/122])$, accounts for elimination of adducts due to the turnover of erythrocytes, with a loss of the oldest fraction of the cells (1/122) each day. Additional factors f may be introduced to account for a decreasing hemoglobin content in aging cells, chemical instability of the adducts, and blood volume changes as described in detail by Osterman-Golkar and Vesper.[13]

When the exposure is long-term ($> t_{er}$) and at a constant level the daily adduct increment α_i could be considered constant $= a$ and when, further, the adducts formed are chemically stable equation (4.1) is usually simplified to:

$$A_{acc} = a \times t_{er}/2 \qquad (4.1b)$$

Scheme 4.1 The basic equation for Hb adduct accumulation.

The turnover of erythrocytes is the main cause of adduct removal. Although most adducts studied to date have been shown to be chemically stable, some adducts, for example those with carboxyl groups, have been shown to be eliminated faster than would be predicted based on erythrocyte turnover. In special cases there might be other parameters that have to be considered in comparisons of Hb adduct levels. In humans, changes in body weight and blood volume may become relevant during pregnancy and in neonates. In experimental animals exposed to carcinogens the dilution of Hb adducts due to increased body weight during the course of the experiment has to be taken into account. Further, exposures at high levels to certain chemicals may cause hemolysis, and increased altitude does increase Hb concentrations. Approximately 20% of the Hb content is lost from the circulating erythrocyte during its life time. This process is most pronounced in old cells and has a marginal effect only on adduct elimination (reviewed by Osterman-Golkar and Vesper[13]). Recent studies have indicated an interindividual variation in the erythrocyte life span. Furne *et al.*[14] found a life span of 122 $\pm$ 23 days in 40 healthy volunteers. It has been believed that fetal red blood cells have a considerably shorter life span than in adults. Later literature however strongly indicates that the fetal erythrocyte life span is about the same as in adults.[15,16]

4.2.3 Acute Single Exposures

Single exposures to reactive compounds are frequently used in experimental animals to study adduct formation and removal. Acute human exposures may occur in connection with accidental release of chemicals, cancer therapy or anaesthesia. In a few studies volunteers have been exposed to defined low concentrations (given in mg/kg body weight) of a carcinogen in order to determine the relation between administered amount (exposure) and *in vivo* dose as measured by Hb adduct levels.[17]

Following a single exposure, chemically stable adducts decline in a nearly linear fashion reaching zero or a background level after a period of time corresponding to the erythrocyte life span. The relatively well characterized and long life span of the erythrocyte is one of the major advantages of Hb adducts as biomarkers. In contrast to adducts with DNA, adducts with Hb are not subjected to repair. Figure 4.3 illustrates the time windows available for measurements of urinary metabolites (typically 1 day) and DNA adducts (the half-life is set to about 4 days in this example) as compared to months in the case of stable Hb adducts.

4.2.4 Long-term Constant Exposures

Adducts with Hb accumulate during prolonged exposure and reach a steady-state level where the rates of adduct formation and adduct removal are equal (Figure 4.4). In the case of chemically stable adducts the steady state is reached

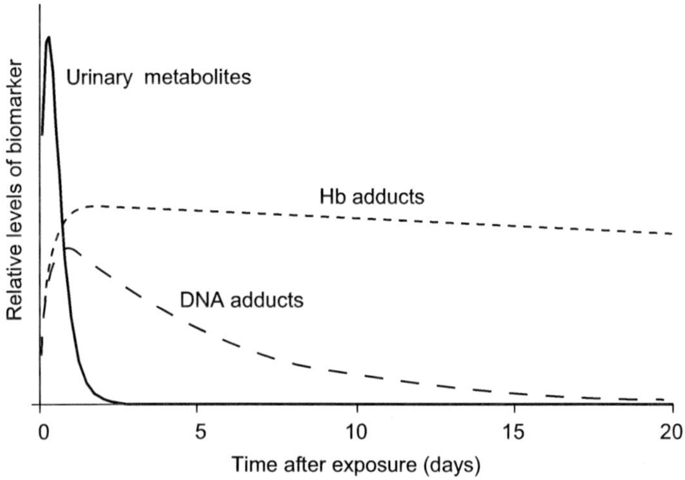

Figure 4.3 Biomarker levels (arbitrary linear scale) after a single high exposure. The Hb adduct level reaches zero (or a background level) after one erythrocyte lifetime.

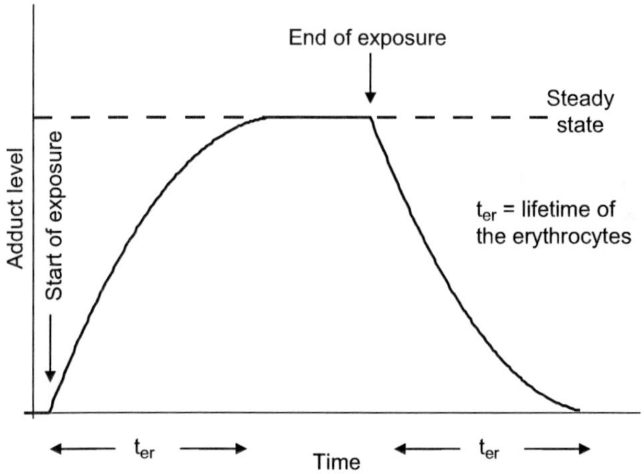

Figure 4.4 Accumulation and removal of chemically stable Hb adducts.

after about 122 days (t_{er}) and the accumulated adduct level could be estimated as about $a \times 61$, where a is the average daily adduct increment [Scheme 4.1, Equation (4.1b)]. Unstable adducts accumulate to a less degree (see Figure 4.4). If the exposure is terminated the adduct levels would decline in a curvilinear manner.[18,19] Environmental and life-style exposures to carcinogenic compounds are generally long term and at a fairly constant level. Thus, measured adduct levels are usually assumed to represent a steady state and could be used for the estimation of the daily average adduct increment and calculation of daily dose [AUC; Scheme 4.1, Equation (4.2)].

4.3 Methodologies for Measurement of Hemoglobin Adducts

The methodologies applied today for the measurement of Hb adducts usually involve mass spectrometry (MS) as the final analytical step. A few studies have employed other detection techniques such as fluorescence detection or immunochemical approaches. Mass spectrometry techniques can offer the high sensitivity and selectivity needed for measurement of the very low levels of adducts that often are encountered. Both gas chromatography (GC) and liquid chromatography (LC) combined with MS are used. GC-MS has been the major technique used, primarily because it has been available for a longer time. Recent developments are often focused on LC-MS techniques. A few applications using capillary electrophoresis in combination with MS have also been described.

Measurement of Hb adducts in the general population from a low background exposure of potentially toxic compounds puts high demands on the analytical performance with regard to sensitivity. N-terminal adducts from, for example, epoxides can be measured down to a few pmol per gram Hb, which corresponds to about 1 modification in 10 million non-modified globin chains. To be able to measure adducts at these low levels the pre-processing of the blood samples is crucial. Which method to choose for processing of the sample and for final detection depends on both the properties of the electrophilic compound of interest and the physical and chemical properties of the adduct formed. Electrophilic compounds differ in reactivity towards different sites in the globin, as mentioned in the Introduction.

There are two principal ways to perform adduct measurements, either by analysis after detachment of the adduct from the amino acid residue in the protein, or by cleavage of peptide bonds and analysis of the modified amino acid or modified peptide. The latter approach, where the analyte includes a moiety from the protein, is advantageous because it increases the specificity of the analyte. Below are examples of frequently used principles.

4.3.1 Detachment of Adducts from the Amino Acid Residue

Aromatic amines, isocyanates and polycyclic aromatic hydrocarbons (PAHs) are examples of compounds that can be analyzed as Hb adducts using mild hydrolysis of the Hb sample. Adducts bound as sulfonamides to cysteine or esters to carboxyl groups can be detached from the nucleophilic atom in the amino acid through alkaline or acidic hydrolysis. The principle was demonstrated about 30 years ago and is applied frequently (reviewed by[20,21]). Different approaches for extraction and enrichment of the free adducts have been described. Liquid–liquid extraction or solid phase extraction has been optimized for different adducts and their specific physical/chemical properties. Derivatization of detached and enriched adducts has been performed to improve chromatographic properties as well as the sensitivity of the analysis. As an example, a methodology frequently applied for biomonitoring of

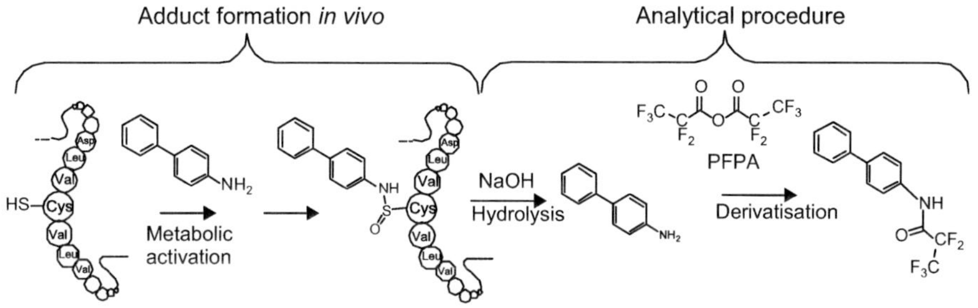

Figure 4.5 Illustration of adduct formation from 4-aminobiphenyl with cysteine and the steps of the analytical procedure. The adduct is detached by a mild hydrolysis with sodium hydroxide (NaOH) and pentafluoropropionic anhydride (PFPA) is used for derivatization to obtain an analyte suitable for GC-MS analysis.

4-aminobiphenyl, a known carcinogen formed in for example cigarette smoke, is illustrated in Figure 4.5. The detached 4-aminobiphenyl is derivatized with pentafluoropropionic anhydride to obtain an analyte with high response when analyzed by GC-MS.

4.3.2 Detachment of Modified Amino Acids by Cleavage of Peptide Bonds

4.3.2.1 Detachment of Modified N-terminal Valines

Adducts with N-terminal valine can be measured with different methodologies, but the principle most frequently applied has been based on the "N-alkyl Edman" procedure. It was first applied for analysis of ethylene oxide Hb adducts but has since then been applied to measurements of several other alkylating agents such as acrylamide, acrylonitrile and several low-molecular-weight epoxides.[8] In contrast to the classical Edman degradation in which an acidification step is required for detachment of N-terminal residues, the adduct-modified N-terminal amino acids detach spontaneously as thiohydantoins at pH $\sim$7 when reacted with phenyl isothiocyanates. A fluorinated Edman reagent has been used for the detachment of the adduct-modified valines, which improves the sensitivity when analysis is carried out with GC-MS (NICI).[22] A high specificity is obtained because the final analyte consists of an intact adduct moiety as well as the amino acid site.[8,22] The properties of the Edman reagent have been exploited to improve different clean-up approaches and detection techniques; for example, fluorescein isothiocyanate has been shown to facilitate analysis with LC-MS.[23,24] The principle is illustrated in Figure 4.6.

4.3.2.2 Enzymatic Digestion (Hydrolysis)

Detachment of adduct-modified amino acid sequences by enzymatic digestion is also a technique with several applications for measurement of adducts with

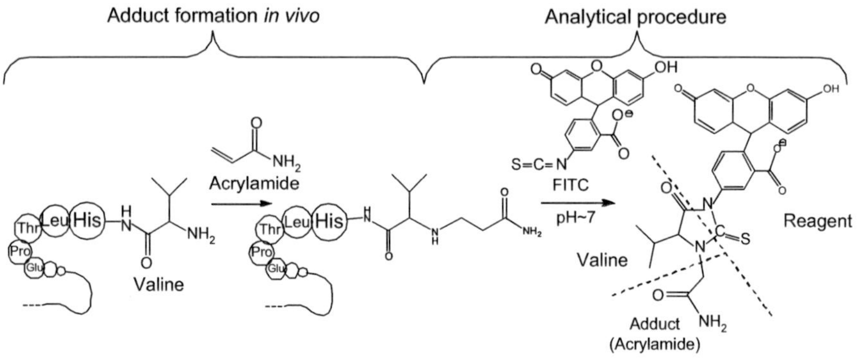

Figure 4.6 Illustration of adduct formation from acrylamide with N-terminal valine and the principles of the analytical procedure when using fluorescein isothiocyanate (FITC) to detach adducted valines. The analyte formed (fluoresceinthiohydantoin) is suitable for analysis with LC-MS. The dashed lines allow visualization of the adduct part, amino acid and reagent part of the thiohydantoin formed.

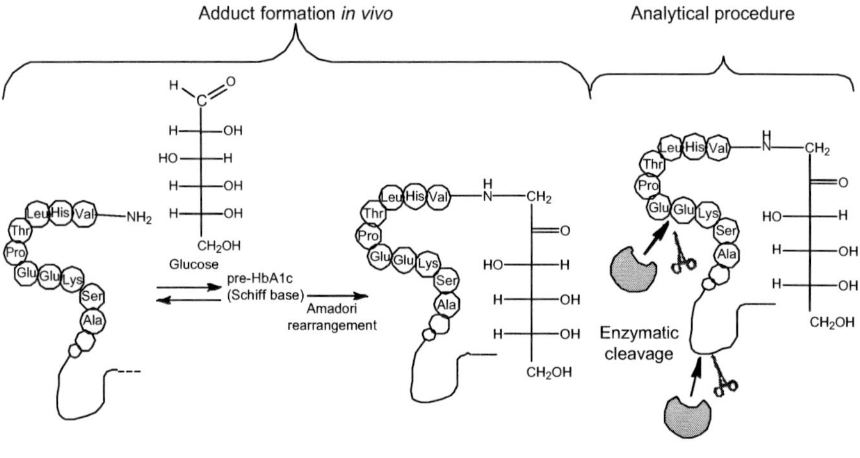

Figure 4.7 Illustration of adduct formation from glucose with N-terminal valine and the principles of the analytical procedure, using the proteolytic enzyme endoproteinase Glu-C. The ratio between the glycated (HbA1c) and the non-glycated ($HbA1_0$) N-terminal hexapeptides of the β-chains is measured. This is used as the reference method recommended by the International Federation of Clinical Chemistry and Laboratory Medicine (IFCC).

Hb as well as with serum albumin. The enzyme usually cleaves at specific sites in the protein chain and the adduct-modified site is left intact. An example is given in Figure 4.7, where the principle of a reference method for measurement of HbA1c is illustrated.[25] Measurement of HbA1c/glycohemoglobin is a widespread application for monitoring of the mean level of glucose in the blood. HbA1c is obtained when glucose forms an adduct with the N-terminal valine of the β-chain in Hb after a two-step non-enzymatic reaction. The carbonyl carbon of the glucose reacts with the amino group on the N-terminal valine and an unstable Schiff base is formed. The Schiff base may either dissociate or undergo an Amadori rearrangement to form a stable adduct. HbA1c is the most abundant fraction (75–80%) in the group of glycohemoglobins that also consists of adducts formed with other sites in Hb and with other monosaccharides. The adducts from glucose in healthy people without diabetes are present in levels as high as a few percentage points of the N-terminal valine in Hb.

4.4 Applications of Hb Adduct Measurement in Human Studies

4.4.1 Occupational Exposures

With regard to human exposure to carcinogens, Hb adducts were initially applied as biomarkers for the detection and quantification of occupational exposure. Several classes of compounds that are reactive or could give reactive metabolites are produced or used as intermediates in the chemical industry. A range of studies has concerned exposure to aromatic amines, which are widely used in the production of dyes, agricultural chemicals, *etc.* Later nitroarenes (also including compounds used as explosives) and aromatic diisocyanates (for production of polyurethane foams and other products) have been biomonitored with the same method (see Figure 4.5).[20,26] Occupational exposure to ethylene oxide, used for instance for sterilization of medical equipment, was early studied by Hb adduct measurement.[6,27] Many studies of exposure to low-molecular-weight electrophilic compounds have followed, using measurement of adducts with N-terminal valines in Hb (*c.f.* Figure 4.6) (reviewed by Ogawa *et al.*[28]). Compounds such as styrene oxide,[29] propylene oxide and epoxides from butadiene,[30,31] acrylonitrile and acrylamide[32] have been monitored in work environments. Isocyanate exposure could be measured specifically as carbamoylated N-termini in Hb, and this method was applied to elucidate the exposure after the Bhopal disaster with high acute exposure to methyl isocyanate (used for pesticide production) in 1984.[33] In workers exposed to dimethylformamide the same adduct (from a metabolite) has been monitored.[34] A class of compounds known as allergens and extensively used in the chemical industry is organic acid anhydrides. For these compounds Hb adducts (released by hydrolysis) could also serve as biomarkers.[35] An influence of genetic polymorphism in metabolizing enzymes on adduct levels has been found in studies of occupational exposure, for instance to acrylonitrile.[36]

Biomonitoring of exposure through Hb adducts has often been of great importance for retrospective assessment of exposure. One illustrative example concerns an acute situation with high exposures to acrylamide in tunnel construction work in Sweden, where Hb adduct measurement was used to clarify the exposure situation. The construction of the railway tunnel through a mountain ridge (Hallandsås) in the south-west of Sweden had encountered difficulties caused by heavy water inflow into the tunnel, and therefore a grouting agent had to be applied to seal the tunnel walls. A product containing the monomers acrylamide and *N*-methylolacrylamide was used in large quantities from August 1997. Subsequently, at the end of September, dead fish in a fish culture and paralyzed cows were found downstream of the rivulet into which leakage water from the tunnel was pumped. The grouting agent was a suspected cause. Hb adducts were measured in the cattle and demonstrated that very high exposure to the acrylamides had occurred (through the drinking water supply). This was taken as proof that the leakage of acrylamides from the tunnel was the cause of the observed effects. Questions were raised immediately about exposure of the tunnel workers and the residents in the area; there could be exposure to acrylamides through inhalation, skin, and even ingestion. The tunnel work was stopped. In the acute phase a few supposedly highly exposed workers and control persons were selected for monitoring of Hb adducts in blood, which was the only possible way to detect and quantify whether any exposure had occurred. The measurements showed adduct levels up to about 4 nmol/g Hb, thus indicating high exposures and that there was a demand for a careful risk assessment. Acrylamide has high reactivity towards proteins and is neurotoxic. It is metabolized to glycidamide which compared to acrylamide has high reactivity towards DNA, and is mutagenic (Figure 4.8). The risk assessment was aided by data generated in a study of acrylamide-exposed workers in China, where Hb adducts had been measured and peripheral nervous system (PNS) symptoms investigated.[32,37] The adduct levels measured in the tunnel workers showed that there was a high risk of PNS symptoms and the assessment of cancer risk found the exposure to be unacceptably high. A complete characterization of the occupational exposure situation for all the workers was done by measurement of Hb adducts (as shown in Figure 4.9).[38] From the results of the examination of PNS symptoms a no-observed adverse effect level (NOAEL) for mild symptoms was estimated and found to correspond to 0.5 nmol/g Hb of the measured acrylamide adduct.[38]

In this example measurement of Hb adducts was not used to monitor an ongoing exposure; it was used to demonstrate and quantify an exposure that had already occurred and to make a risk assessment. In other cases

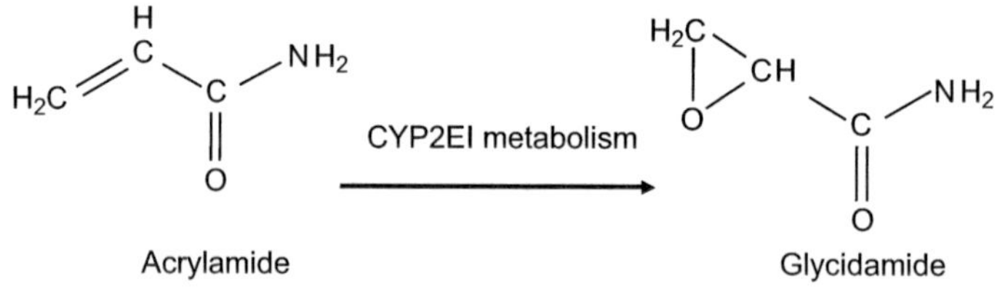

Figure 4.8 Acrylamide is metabolized to the mutagenic epoxide glycidamide by cytochrome P450 2EI.

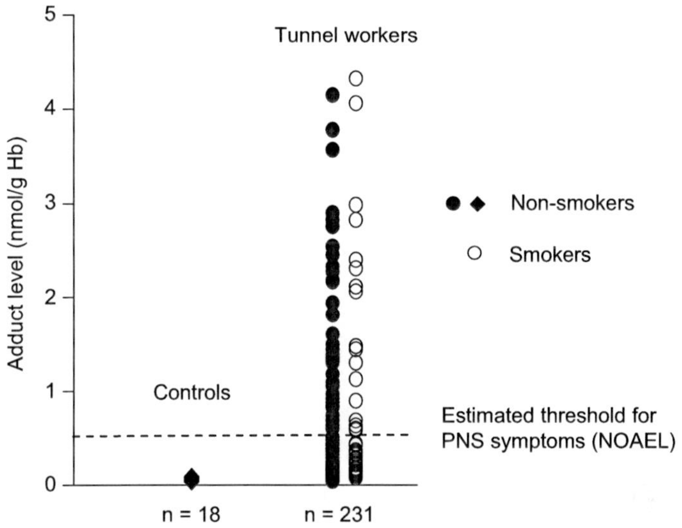

Figure 4.9 Hb adducts from acrylamide measured in potentially exposed tunnel workers and controls. From the results of neurophysiological examination a threshold (no-observed adverse effect level, NOAEL) of 0.5 nmol/g was estimated for mild, reversible symptoms of the peripheral nervous system (PNS). (From Hagmar *et al.*[38])

measurement of Hb adducts has been used to control or improve the work environment and/or to exclude exposure and thus reduce anxiety.[39]

4.4.2 Clinical Applications

There is one specific application of Hb adduct measurement that is now used as a routine method. This is the measurement of glycated Hb (see Figure 4.7), which is used in the assessment of glycemic control in patients with diabetes and which can provide an indication of how well diabetes is being managed. A similar, less frequently used application is the monitoring of urea levels associated with renal failure by measurement of increased carbamoylation of N-terminal valine in Hb.[40]

With regard to exposure to genotoxic compounds there are a limited number of clinical applications of Hb adduct measurement so far. One example concerns the widely used drug lidocaine, which has been shown to form genotoxic metabolites of dimethylaniline, measured as Hb adducts, in treated patients.[41] Monitoring of Hb adducts has been tested for use in monitoring of chemotherapy by nitrosoureas.[42] A few cases are described in the literature in which Hb adducts have been used to identify exposure to sulfur mustard in victims of the Iran–Iraq conflict.[43,44] Another illustration of application within medicine is a case in which analysis of acrylamide adducts was used to aid in the diagnosis of severe ill-health in one person, revealing that unknown high exposure to acrylamide was the most probable cause of the observed symptoms.[39]

4.4.3 Environmental Exposures

Adducts from aromatic amines have been measured in many studies of smokers and non-smokers (reviewed by Sabbioni and Jones[20]). It has been shown that adduct levels from several aromatic amines are increased in smokers compared with non-smokers. In particular, adducts from 3- and 4-aminobiphenyl (ABP) have been studied in depth. For instance, it has been shown that the levels of 4-ABP adducts increase with exposure to environmental tobacco smoke. The level of 4-ABP adducts in non-smokers is consistent with contamination in air, food and water. Individual susceptibilities with regard to metabolic enzymes have been studied, and it has been shown that "slow" acetylators have a higher adduct level from 3- and 4-ABP than "rapid" acetylators at similar exposure levels.

Urban air pollution has been one subject of study, and ethylene, which is metabolized to ethylene oxide, is one air pollutant studied in depth. Also in this case smokers have an increased level of adducts. Studies of the origin of the background levels (*ca.* 20 pmol/g Hb) of this adduct in non-smokers showed that the major contribution arose from endogenous ethylene/ethylene oxide (reviewed by Törnqvist and Kautianen[45]), which thus makes it difficult to observe an increment caused by exposure from air pollution. Furthermore, animal experiments showed an influence on the adduct level from fat in the diet and intestinal flora.

The results from studies of adducts from aromatic amines and ethylene oxide demonstrate clearly that Hb adduct measurements could detect an exposure to potentially genotoxic compounds due to complex exposure from tobacco smoking. More interesting, though, is the fact that background exposure in non-smokers from environmental exposure, but also from endogenous production, has been demonstrated with measurements of Hb adducts.

The investigation of the exposure to acrylamide in the tunnel construction work (see above) initiated studies that disclosed that acrylamide is present in many staple foods. A background level of acrylamide adducts with Hb from non-smoking control persons (*ca.* 40 pmol/g; see Figure 4.9) had already been indicated earlier.[46] If the observation in non-smokers really originated from a background exposure to acrylamide it would correspond to a rather high continuous exposure (daily intakes of 10–100 μg). The cancer risk evaluations done in the context of the exposed tunnel workers strongly indicated that this potential background exposure had to be studied in more depth. It was hypothesized that the adduct level in non-smokers was related to heated food. This was based on the fact that acrylamide adduct levels are higher in smokers, and on the observation of very low adduct levels in different animal species studied as controls in the context of the acrylamide leakage during construction of the tunnel. Experiments showed that rats given fried laboratory animal feed had clearly much higher acrylamide adduct levels than the control rats. The identity of the adduct was proven by mass spectrometric methods and higher levels of acrylamide were also demonstrated in the fried feed.[47] In the next step researchers studied whether acrylamide could be formed in human foodstuffs, and high levels of acrylamide were demonstrated in potato products, bread, *etc.* during normal preparation at high temperatures.[48]

The finding that acrylamide is formed in foods was unexpected, and led to concerns about health risks in the population associated with the exposure to acrylamide via food. Research was initiated in related fields, and biomonitoring of Hb adducts from acrylamide has since then been applied in many experimental and molecular epidemiological studies. Table 4.1 shows background levels of the acrylamide Hb adduct (to N-termini) measured in non-smokers from different populations. The relatively small variations in median levels, 25–50 pmol/g, observed in these studies reflect variations both between methods/laboratories and among populations.[49–63] Small differences have, however, been observed between populations from different countries, which are most likely to be explained by eating habits.[59] Measured Hb adduct levels also have indicated a higher intake of acrylamide in adolescents and children.[57,63] The presence of acrylamide in foods has also been verified in innumerable studies. Mean daily intakes have been estimated to be 0.5–1.0 μg/kg body weight (35–70 μg/person).

A human study that involved high intake of acrylamide-rich foods, with measurement of acrylamide concentrations, for 4 days showed clear-cut relationships between intake and the increments in adduct levels for both

Table 4.1 Levels of adducts from acrylamide to N-termini measured in hemoglobin from non-smokers. Some studies include the same study subjects, as indicated in the table. Both men and women were tested if not otherwise indicated.

Adduct level (pmol/g globin)			
Median	*Range*	*n*	*Reference*
40/40	16–100	35 (men)/	49
		35 (women)	50
28 ± 7 Mean ($\pm$SD)		100	51,52
31/22	10–70	21 (men)/	53
	9–50	31 (women)	
37	18–66	44	54
51	7–610	73	55
26	3–103	857	56,57
35	17–96	484 (women)	58
43/42	15–177	120 (men)/	59
		135 (women)	
47	16–179	331 (men)	60,61
44	14–148	296 (women)	62
50/51	3– ~500	3050 (women)/	63
		2636 (men)	

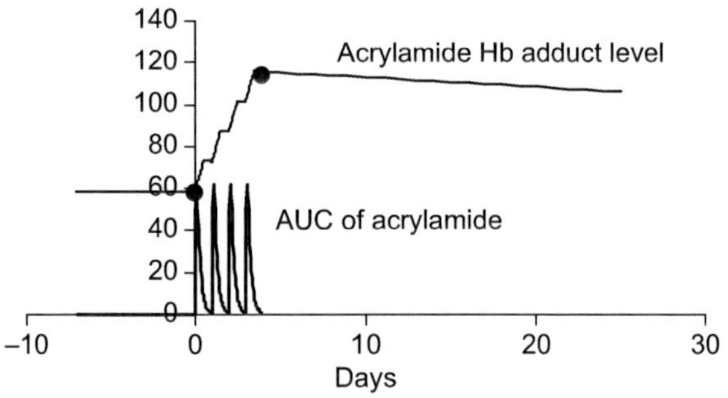

Figure 4.10 Nine non-smokers were maintained on a carefully monitored acrylamide-rich diet for 4 days. The average daily intake was *ca.* 11 µg acrylamide/kg body weight. The hypothetical bottom curve illustrates the uptake and elimination of acrylamide assuming that the uptake occurs in the middle of the day and assuming a half-life of acrylamide of 3 hours. The circles show the adduct levels measured immediately before and after exposure period (pmol/g Hb). The full line shows a simulation of adduct formation and removal during and after the dietary exposure. (Data from Vikström *et al.*[64])

acrylamide and the metabolite glycidamide.[64] Figure 4.10 illustrates the daily increment in adduct level above the present background level of the adduct.[64] The lower curve illustrates the "area under the concentration curve" from the intake of acrylamide. Such data from intervention experiments could be used to calculate the mean intakes from adduct levels measured in groups.

Hb adducts from ethylene oxide, propylene oxide, methylating and ethylating agents, acrylonitrile or acrylamide have been measured in several studies of smokers and non-smokers. Higher levels have been found in smokers throughout, and background levels have been found in non-smokers (for instance by ref. 52). For some compounds a correlation of adduct level with genotypes for metabolizing enzymes has been observed, for instance for glutathione transferase T1 and adduct levels from ethylene oxide.[65]

4.4.4 Limitations

Monitoring of Hb adducts has been shown to be very useful for many simple alkylating agents, as illustrated above. Limitations are indicated; for instance bulky compounds may have difficulty in crossing the erythrocyte membrane. Hb adducts, however, at very low levels, have been measured from PAH (reviewed by Kafferlein *et al.*[66]). Measurement of adducts to serum albumin is an alternative, and have been applied successfully, for instance for measurement of high exposures to benzene.[67] Another example is exposure to aflatoxin, which has been monitored through serum albumin adducts in several populations.[68]

4.5 Future Perspectives

The use of protein adduct monitoring is expected to increase with regard to the studied compounds and monitor proteins used. One example is recent studies that demonstrate that certain beneficial exposures could be monitored through adducts. This concerns isothiocyanates, which have been shown to have cancer preventive effects in animal experiments.[69] Other recent developments include use of adducts with keratin to measure dermal exposure specifically.[70]

An interesting area for further research is the exploration of the significance of background adducts as a reflection of exposure from exogenous and endogenous sources of electrophilic compounds and of metabolism. Several methods for adduct analysis are developed to semi-high throughput methods,[71,72] and it is now possible to measure adducts in large cohorts. The largest cohorts studied concern monitoring of the intake of acrylamide from food in the general population,[56,63] and mother–child cohorts (von Stedingk *et al.*, to be published) with more than 1000 blood samples analyzed. Methodological developments also strive towards analysis of small volumes of blood.[63,73] The new developments could facilitate the recent adductomic approach, which is aimed at the identification of unknown adducts or finding adduct patterns that signify health status or exposures.

Acknowledgements

Economic support from the Swedish Cancer and Allergy Foundation, Swedish Research Council Formas and the EU Integrated Project NewGeneris, 6th Framework Programme, Priority 5: Food Quality and Safety (Contract no. FOOD-CT-2005-016320) is acknowledged.

References

1. H. Jackson and R. Thompson, The reactions of some haemoglobin derivatives with 131I-labelled p-iodophenylhydroxylamine and p-iodonitrosobenzene, *Biochem. J.*, 1953, **55**, 36–37.
2. J. Lewalter and U. Korallus, Blood protein conjugates and acetylation of aromatic amines. New findings on biological monitoring, *Int. Arch. Occup. Environ. Health,* 1985, **56**, 179–196.
3. H. G. Neumann, Analysis of hemoglobin as a dose monitor for alkylating and arylating agents, *Arch. Toxicol.*, 1984, **56**, 1–6.
4. L. Ehrenberg, Genetic toxicity of environmental chemicals, *Acta Biol. Yugosl. Ser. F: Genetika*, 1974, **6**, 367–398.
5. S. Osterman-Golkar, L. Ehrenberg, D. Segerbäck and I. Hällstrom, Evaluation of genetic risk of alkylating agents. II. Haemoglobin as dose monitor, *Mutat. Res.*, 1976, **34**, 1–10.
6. C. J. Calleman, L. Ehrenberg, B. Jansson, S. Osterman-Golkar, D. Segerback, K. Svensson and C. A. Wachtmeister, Monitoring and risk assessment by means of alkyl groups in hemoglobin in persons

occupationally exposed to ethylene oxide, *J. Environ. Pathol. Toxicol.*, 1978, **2**, 427–442.
7. C. G. Swain and C. B. Scott, Quantitative correlation of relative rates. Comparison of hydroxide ion with other nucleophilic reagents towards alkyl halides, esters, epoxides and acyl halides, *J. Am. Chem. Soc.*, 1953, **75**, 141–147.
8. M. Törnqvist, C. Fred, J. Haglund, H. Helleberg, B. Paulsson and P. Rydberg, Protein adducts: quantitative and qualitative aspects of their formation, analysis and applications, *J. Chromatogr. B*, 2002, **778**, 279–308.
9. P. L. Skipper and S. R. Tannenbaum, Protein adducts in the molecular dosimetry of chemical carcinogens, *Carcinogenesis*, 1990, **11**, 507–518.
10. F. N. Granath, C. E. Vaca, L. G. Ehrenberg and M. A. Törnqvist, Cancer risk estimation of genotoxic chemicals based on target dose and a multiplicative model, *Risk Anal.*, 1999, **19**, 309–320.
11. D. Segerbäck, Reaction products in hemoglobin and DNA after *in vitro* treatment with ethylene oxide and N-(2-hydroxyethyl)-N-nitrosourea, *Carcinogenesis*, 1990, **11**, 307–312.
12. T. R. Fennell, J. P. MacNeela, M. J. Turner and A. Swenberg, *Hemoglobin Adduct Formation by Acrylonitrile in Rats and Mice*, Oxford University Press, 1991, pp. 241–246.
13. S. M. Osterman-Golkar and H.W. Vesper, Assessment of the relationship between glucose and A1c using kinetic modeling, *J. Diabetes Complications*, 2006, **20**, 285–294.
14. J. K. Furne, J. R. Springfield, S. B. Ho and M. D. Levitt, Simplification of the end-alveolar carbon monoxide technique to assess erythrocyte survival, *J. Lab. Clin. Med.*, 2003, **142**, 52–57.
15. R. A. Brace, C. Langendorfer, T. B. Song and D. M. Mock, Red blood cell life span in the ovine fetus, *Am. J. Physiol. Regul. Integr. Comp. Physiol.*, 2000, **279**, R1196–R1204.
16. M. H. Dziegiel, O. Koldkjaer and A. Berkowicz, Massive antenatal fetomaternal hemorrhage: evidence for long-term survival of fetal red blood cells, *Transfusion*, 2005, **45**, 539–544.
17. T. R. Fennell, S. C. Sumner, R. W. Snyder, J. Burgess, R. Spicer, W. E. Bridson and M. A. Friedman, Metabolism and hemoglobin adduct formation of acrylamide in humans, *Toxicol. Sci.*, 2005, **85**, 447–459.
18. F. Granath, L. Ehrenberg and M. Tornqvist, Degree of alkylation of macromolecules *in vivo* from variable exposure, *Mutat. Res.*, 1992, **284**, 297–306.
19. T. R. Fennell, S. C. Sumner and V. E. Walker, A model for the formation and removal of hemoglobin adducts, *Cancer Epidemiol. Biomarkers Prev.*, 1992, **1**, 213–219.
20. G. Sabbioni and C. R. Jones, Biomonitoring of arylamines and nitroarenes, *Biomarkers*, 2002, **7**, 347–421.
21. E. Richter and B. Branner, Biomonitoring of exposure to aromatic amines: haemoglobin adducts in humans, *J. Chromatogr. B*, 2002, **778**, 49–62.
22. M. Tornqvist, J. Mowrer, S. Jensen and L. Ehrenberg, Monitoring of environmental cancer initiators through hemoglobin adducts by a modified Edman degradation method, *Anal. Biochem.*, 1986, **154**, 255–266.

23. P. Rydberg, H. von Stedingk, J. Magner and J. Bjorklund, LC/MS/MS analysis of N-terminal protein adducts with improved sensitivity: a comparison of selected Edman isothiocyanate reagents, *Int. J. Anal. Chem.*, 2009, **2009**, 153472.
24. P. Rydberg, Method for analyzing N-terminal protein adducts using isothiocyanate reagents, *PCT Int. Appl.*, 2005, WO2005101020, 63.
25. J. O. Jeppsson, U. Kobold, J. Barr, A. Finke, W. Hoelzel, T. Hoshino, K. Miedema, A. Mosca, P. Mauri, R. Paroni, L. Thienpont, M. Umemoto and C. Weykamp, Approved IFCC reference method for the measurement of HbA1c in human blood, *Clin. Chem. Lab. Med.*, 2002, **40**, 78–89.
26. G. Sabbioni, H. Wesp, J. Lewalter and R. Rumler, Determination of isocyanate biomarkers in construction site workers, *Biomarkers*, 2007, **12**, 468–483.
27. P. B. Farmer, E. Bailey, S. M. Gorf, M. Tornqvist, S. Osterman-Golkar, A. Kautiainen and D. P. Lewis-Enright, Monitoring human exposure to ethylene oxide by the determination of haemoglobin adducts using gas chromatography-mass spectrometry, *Carcinogenesis,* 1986, **7**, 637–640.
28. M. Ogawa, T. Oyama, T. Isse, T. Yamaguchi, T. Murakami, Y. Endo and T. Kawamoto, Hemoglobin adducts as a marker of exposure to chemical substances, especially PRTR class I designated chemical substances, *J. Occup. Health*, 2006, **48**, 314–328.
29. J. P. Teixeira, J. Gaspar, J. Roma-Torres, S. Silva, C. Costa, J. Roach, O. Mayan, J. Rueff and P. B. Farmer, Styrene-oxide N-terminal valine haemoglobin adducts in reinforced plastic workers: possible influence of genetic polymorphism of drug-metabolising enzymes, *Toxicology*, 2007, **237**, 58–64.
30. P. J. Boogaard, Use of haemoglobin adducts in exposure monitoring and risk assessment, *J. Chromatogr. B, Anal. Technol. Biomed. Life Sci.*, 2002, **778**, 309–322.
31. P. M. Vacek, R. J. Albertini, R. J. Sram, P. Upton and J. A. Swenberg, Hemoglobin adducts in 1,3-butadiene exposed Czech workers: female–male comparisons, *Chem. Biol. Interact.*, 2010, **188**, 668–676.
32. E. Bergmark, C. J. Calleman, F. He and L. G. Costa, Determination of hemoglobin adducts in humans occupationally exposed to acrylamide, *Toxicol. Appl. Pharmacol.*, 1993, **120**, 45–54.
33. S. Sriramachari, G. J. Rao, V. K. Sharma, R. K. Jadhav, A. K. Saraf and H. Chandra, GC-NPD and GC-MS analysis of preserved tissue of Bhopal gas disaster: evidence of methyl carbamylation in post-mortem blood, *Med. Sci. Law,* 1991, **31**, 289–293.
34. H. U. Kafferlein and J. Angerer, N-methylcarbamoylated valine of hemoglobin in humans after exposure to N,N-dimethylformamide: evidence for the formation of methyl isocyanate?, *Chem. Res. Toxicol.*, 2001, **14**, 833–840.
35. S. Rosqvist, G. Johannesson, C. H. Lindh and B. A. Jonsson, Total plasma protein adducts of allergenic hexahydrophthalic and methylhexahydrophthalic anhydrides as biomarkers of long-term exposure, *Scand. J. Work Environ. Health*, 2001, **27**, 133–139.
36. R. Thier, H. Balkenhol, J. Lewalter, S. Selinski, A. Dommermuth and H. M. Bolt, Influence of polymorphisms of the human glutathione

transferases and cytochrome P450 2E1 enzyme on the metabolism and toxicity of ethylene oxide and acrylonitrile, *Mutat. Res.*, 2001, **482**, 41–46.
37. C. J. Calleman, Y. Wu, F. He, G. Tian, E. Bergmark, S. Zhang, H. Deng, Y. Wang, K. M. Crofton and T. Fennell, Relationships between biomarkers of exposure and neurological effects in a group of workers exposed to acrylamide, *Toxicol. Appl. Pharmacol.*, 1994, **126**, 361–371.
38. L. Hagmar, M. Törnqvist, C. Nordander, I. Rosén, M. Bruze, A. Kautiainen, A.-L. Magnusson, B. Malmberg, P. Aprea, F. Granath and A. Axmon, Health effects of occupational exposure to acrylamide using hemoglobin adducts as biomarkers of internal dose, *Scand. J. Work Environ. Health*, 2001, **27**, 219–226.
39. B. Paulsson, K.-O. Larsen and M. Törnqvist, Hemoglobin adducts in the assessment of potential occupational exposure to acrylamides – three case studies, *Scand. J. Work Environ. Health*, 2006, **32**, 154–159.
40. A. Wynckel, C. Randoux, H. Millart, C. Desroches, P. Gillery, E. Canivet and J. Chanard, Kinetics of carbamylated haemoglobin in acute renal failure, *Nephrol. Dial. Transplant*, 2000, **15**, 1183–1188.
41. M. S. Bryant, H. F. Simmons, R. E. Harrell and J. A. Hinson, 2,6-Dimethylaniline–hemoglobin adducts from lidocaine in humans, *Carcinogenesis*, 1994, **15**, 2287–2290.
42. E. Bailey, P. B. Farmer, Y. S. Tang, H. Vangikar, A. Gray, D. Slee, R. M. Ings, D. B. Campbell, J. G. McVie and R. Dubbelman, Hydroxyethylation of hemoglobin by 1-(2-chloroethyl)-1-nitrosoureas, *Chem. Res. Toxicol.*, 1991, **4**, 462–466.
43. H. P. Benschop, G. P. van der Schans, D. Noort, A. Fidder, R. H. Mars-Groenendijk and L. P. de Jong, Verification of exposure to sulfur mustard in two casualties of the Iran-Iraq conflict, *J. Anal. Toxicol.*, 1997, **21**, 249–251.
44. R. M. Black, R. J. Clarke, J. M. Harrison and R. W. Read, Biological fate of sulphur mustard: identification of valine and histidine adducts in haemoglobin from casualties of sulphur mustard poisoning, *Xenobiotica*, 1997, **27**, 499–512.
45. M. Tornqvist and A. Kautiainen, Adducted proteins for identification of endogenous electrophiles, *Environ. Health Perspect.*, 1993, **99**, 39–44.
46. E. Bergmark, Hemoglobin adducts of acrylamide and acrylonitrile in laboratory workers, smokers and nonsmokers, *Chem. Res. Toxicol.*, 1997, **10**, 78–84.
47. E. Tareke, P. Rydberg, P. Karlsson, S. Eriksson and M. Tornqvist, Acrylamide: a cooking carcinogen?, *Chem. Res. Toxicol.*, 2000, **13**, 517–522.
48. E. Tareke, P. Rydberg, P. Karlsson, S. Eriksson and M. Tornqvist, Analysis of acrylamide, a carcinogen formed in heated foodstuffs, *J. Agric. Food Chem.*, 2002, **50**, 4998–5006.
49. L. Hagmar, E. Wirfält, B. Paulsson and M. Törnqvist, Differences in hemoglobin adduct levels of acrylamide in the general population with respect to dietary intake, smoking habits and gender, *Mutat. Res.*, 2005, **580**, 157–165.
50. E. Wirfalt, B. Paulsson, M. Tornqvist, A. Axmon and L. Hagmar, Associations between estimated acrylamide intakes and hemoglobin AA adducts in a sample from the Malmo Diet and Cancer cohort, *Eur. J. Clin. Nutr.*, 2008, **62**, 314–323.

51. M. Urban, D. Kavvadias, K. Riedel, G. Scherer and A.R. Tricker, Urinary mercapturic acids and a hemoglobin adduct for the dosimetry of acrylamide exposure in smokers and nonsmokers, *Inhal. Toxicol.*, 2006, **18**, 831–839.
52. G. Scherer, J. Engl, M. Urban, G. Gilch, D. Janket and K. Riedel, Relationship between machine-derived smoke yields and biomarkers in cigarette smokers in Germany, *Regul. Toxicol. Pharmacol.*, 2007, **47**, 171–183.
53. S. Chevolleau, C. Jacques, C. Canlet, J. Tulliez and L. Debrauwer, Analysis of hemoglobin adducts of acrylamide and glycidamide by liquid chromatography-electrospray ionization tandem mass spectrometry, as exposure biomarkers in French population, *J. Chromatogr. A*, 2007, **1167**, 125–134.
54. T. Bjellaas, P. T. Olesen, H. Frandsen, M. Haugen, L. H. Stoelen, J. E. Paulsen, J. Alexander, E. Lundanes and G. Becher, Comparison of estimated dietary intake of acrylamide with hemoglobin adducts of acrylamide and glycidamide, *Toxicol. Sci.*, 2007, **98**, 110–117.
55. H. W. Vesper, J. T. Bernert, M. Ospina, T. Meyers, L. Ingham, A. Smith and G. L. Myers, Assessment of the relation between biomarkers for smoking and biomarkers for acrylamide exposure in humans, *Cancer Epidemiol. Biomarkers Prev.*, 2007, **16**, 2471–2478.
56. B. Kutting, W. Uter and H. Drexler, The association between self-reported acrylamide intake and hemoglobin adducts as biomarkers of exposure, *Cancer Causes Control*, 2008, **19**, 273–281.
57. B. Kutting, T. Schettgen, U. Schwegler, H. Fromme, W. Uter, J. Angerer and H. Drexler, Acrylamide as environmental noxious agent: a health risk assessment for the general population based on the internal acrylamide burden, *J. Hyg. Environ. Health,* 2009, **212**, 470–480.
58. P. T. Olesen, A. Olsen, H. Frandsen, K. Frederiksen, K. Overvad and A. Tjoenneland, Acrylamide exposure and incidence of breast cancer among postmenopausal women in the Danish Diet, Cancer and Health study, *Int. J. Cancer*, 2008, **122**, 2094–2100.
59. H. W. Vesper, N. Slimani, G. Hallmans, A. Tjonneland, A. Agudo, V. Benetou, S. Bingham, H. Boeing, M. C. Boutron-Ruault, H. B. Bueno-de-Mesquita, D. Chirlaque, F. Clavel-Chapelon, F. Crowe, D. Drogan, P. Ferrari, I. Johansson, R. Kaaks, J. Linseisen, E. Lund, J. Manjer, A. Mattiello, D. Palli, P. H. Peeters, S. Rinaldi, G. Skeie, A. Trichopoulou, P. Vineis, E. Wirfalt, K. Overvad and U. Stromberg, Cross-sectional study on acrylamide hemoglobin adducts in subpopulations from the European Prospective Investigation into Cancer and Nutrition (EPIC) Study, *J. Agric. Food Chem.*, 2008, **56**, 6046–6053.
60. K. M. Wilson, K. Balter, H. O. Adami, H. Gronberg, A. C. Vikstrom, B. Paulsson, M. Tornqvist and L. A. Mucci, Acrylamide exposure measured by food frequency questionnaire and hemoglobin adduct levels and prostate cancer risk in the Cancer of the Prostate in Sweden Study, *Int. J. Cancer*, 2009, **124**, 2384–2390.
61. A. C. Vikstrom, K. M. Wilson, B. Paulsson, I. Athanassiadis, H. Gronberg, H. O. Adami, J. Adolfsson, L. A. Mucci, K. Balter and M. Tornqvist, Alcohol

influence on acrylamide to glycidamide metabolism assessed with hemoglobin-adducts and questionnaire data, *Food Chem. Toxicol.*, 2010, **48**, 820–824.
62. K. M. Wilson, H. W. Vesper, P. Tocco, L. Sampson, J. Rosen, K. E. Hellenas, M. Tornqvist and W. C. Willett, Validation of a food frequency questionnaire measurement of dietary acrylamide intake using hemoglobin adducts of acrylamide and glycidamide, *Cancer Causes Control*, 2009, **20**, 269–278.
63. H. W. Vesper, S. P. Caudill, J. D. Osterloh, T. Meyers, D. Scott and G. L. Myers, Exposure of the U.S. population to acrylamide in the National Health and Nutrition Examination Survey 2003–2004, *Environ. Health Perspect.*, 2010, **118**, 278–283.
64. A. C. Vikstrom, L. Abramsson-Zetterberg, M. Naruszewicz, I. Athanassiadis, F. N. Granath and M. A. Tornqvist, *In vivo* doses of acrylamide and glycidamide in humans after intake of acrylamide-rich food, *Toxicol. Sci.*, 2011, **119**, 41–49.
65. T. R. Fennell, J. P. MacNeela, R. W. Morris, M. Watson, C. L. Thompson and D. A. Bell, Hemoglobin adducts from acrylonitrile and ethylene oxide in cigarette smokers: effects of glutathione S-transferase T1-null and M1-null genotypes, *Cancer Epidemiol. Biomarkers Prev.*, 2000, **9**, 705–712.
66. H. U. Kafferlein, B. Marczynski, T. Mensing and T. Bruning, Albumin and hemoglobin adducts of benzo[a]pyrene in humans – analytical methods, exposure assessment, and recommendations for future directions, *Crit. Rev. Toxicol.*, 2010, **40**, 126–150.
67. Y. S. Lin, P. P. Egeghy and S. M. Rappaport, Relationships between levels of volatile organic compounds in air and blood from the general population, *J. Expo. Sci. Environ. Epidemiol*, 2008, **18**, 421–429.
68. L. F. McCoy, P. F. Scholl, A. E. Sutcliffe, S. M. Kieszak, C. D. Powers, H. S. Rogers, Y. Y. Gong, J. D. Groopman, C. P. Wild and R. L. Schleicher, Human aflatoxin albumin adducts quantitatively compared by ELISA, HPLC with fluorescence detection, and HPLC with isotope dilution mass spectrometry, *Cancer Epidemiol. Biomarkers Prev.*, 2008, **17**, 1653–1657.
69. A. Kumar, P. Vineis, C. Sacerdote, L. Fiorini and G. Sabbioni, Determination of new biomarkers to monitor the dietary consumption of isothiocyanates, *Biomarkers*, 2010, **15**, 739–745.
70. J. C. Kang-Sickel, V. P. Stober, J. E. French and L. A. Nylander-French, Exposure to naphthalene induces naphthyl-keratin adducts in human epidermis *in vitro* and *in vivo*, *Biomarkers*, 2010, **15**, 488–497.
71. H. von Stedingk, P. Rydberg and M. Tornqvist, A new modified Edman procedure for analysis of N-terminal valine adducts in hemoglobin by LC-MS/MS, *J. Chromatogr. B*, 2010, **878**, 2483–2490.
72. H. W. Vesper, M. Ospina, T. Meyers, L. Ingham, A. Smith, J. G. Gray and G. L. Myers, Automated method for measuring globin adducts of acrylamide and glycidamide at optimized Edman reaction conditions, *Rapid Commun. Mass Spectrom.*, 2006, **20**, 959–964.
73. W. E. Funk, S. Waidyanatha, S. H. Chaing and S. M. Rappaport, Hemoglobin adducts of benzene oxide in neonatal and adult dried blood spots, *Cancer Epidemiol. Biomarkers Prev.*, 2008, **17**, 1896–1901.

CHAPTER 5

Biomarkers of Human Exposure to Environmental Tobacco Smoke (ETS)

DAVID H. PHILLIPS

Institute of Cancer Research, Brookes Lawley Building, Cotswold Road, Sutton, Surrey SM2 5NG, UK

5.1 The Ill-health Effects of Tobacco Smoking – Active and Passive

It is beyond dispute that cigarette smoking is a leading cause of premature death and of preventable disease in many societies. Prominent among the harmful effects of smoking are cardiovascular disease, cancer and pulmonary disease, but smokers are also at greater risk than non-smokers of many other conditions, including respiratory tract infections, osteoporosis, reproductive disorders, duodenal and gastric ulcers and diabetes.[1] In Western countries, smoking is a contributory factor in 1 in 5 deaths and it is estimated that 50% of lifelong smokers die prematurely of a smoking-related disease; on average, persistent smokers die 10 years younger than non-smokers.[2]

More than 50 years after the landmark report of Doll and Hill[3] on mortality among British doctors due to lung cancer from smoking, new evidence of the harm to health of active smoking continues to accumulate. When first evaluated by the International Agency for Research on Cancer (IARC) in 1985, tobacco smoking was concluded to cause cancers of the lung, bladder, renal

Issues in Toxicology No. 10
Biomarkers and Human Biomonitoring, Volume 2: Selected Biomarkers of Current Interest
Edited by Lisbeth E. Knudsen and Domenico Franco Merlo

Published by the Royal Society of Chemistry, www.rsc.org

pelvis, oral cavity (including the lip), oropharynx, hypopharynx, larynx, oesophagus and pancreas.[4] When tobacco smoking was re-evaluated in 2002, cancers of the ureter, stomach, liver, uterine cervix, tongue, nasal cavity and paranasal sinuses, nasopharynx and bone marrow (myeloid leukaemia) were added to the list,[5] followed in 2009 by cancers of the colorectum and ovary, with limited evidence for an association with cancer of the female breast[6]. In fact currently the only tumour sites for which there is evidence suggesting a lack of association between smoking and carcinogenicity are endometrium and thyroid.[6]

More recently, attention has turned to the effects of tobacco smoke on non-smokers. The inhalation of tobacco smoke by non-smokers is referred to by a number of terms, including exposure to secondhand smoke; exposure to environmental tobacco smoke (ETS); passive smoking; and involuntary smoking. These terms are used interchangeably in this chapter.

Many studies have reported that exposure to secondhand smoke is a risk factor for coronary heart disease.[7] For reasons that are not entirely clear, the risk appears to be disproportionately high relative to the risk from active smoking, for which the dose–response curve is non-linear, increasingly rapidly at low doses but showing relatively little further increase at higher doses.[8] Recent estimates are that secondhand smoke causes over 35 000 deaths from coronary heart disease in the United States.[9] There are also reports of a higher risk of stroke among non-smokers exposed to secondhand smoke, relative to those not exposed.[10] Taken together, these characteristics suggest that the health benefits to the general population from increased restriction on smoking in public places may be dramatic,[8] and there is now considerable evidence that implementation of comprehensive smoke-free legislation leads to a reduction in acute coronary events.[11]

The IARC evaluations first considered the risk of involuntary smoking in 2002, when it was concluded that there was sufficient evidence that secondhand smoke causes lung cancer.[5] This was re-affirmed in 2009, when it was also concluded that there is now limited evidence for an association with cancer of the larynx and pharynx.[6] In contrast to the risk to non-smokers of acute myocardial infarction, the excess risk for lung cancer from exposure to secondhand smoke follows a more linear dose–response relationship. It should be borne in mind that even though lung cancer is much less common in non-smokers than in smokers, it is nevertheless still a leading cause of cancer mortality among non-smokers.[12] Primary causative factors include radon and indoor air pollution, in addition to secondhand tobacco smoke, although the origin of a large fraction of lung cancer in people who have never smoked remains poorly understood.[13]

Exposure of infants and children to ETS, often from parental smoking in the home, is associated with many adverse health effects, including sudden infant death syndrome (SIDS),[14,15] asthma, middle ear infection (otitis media) and lower respiratory tract illness.[16,17] Prenatal exposure to tobacco smoke also carries health risks, with adverse effects on fertility, conception and survival of the conceptus.[18] It is well documented that smoking by mothers during pregnancy impairs fetal growth. The hypothesis that maternal smoking also harms the developing fetal brain is supported by an association of increased psychiatric morbidity and exposure to tobacco smoking *in utero* and, where more than 10 cigarettes per day were smoked during pregnancy, there was a greater risk of mortality in childhood, adolescence and young adulthood.[19] Prenatal exposure to tobacco smoke is also a risk factor for SIDS.[14,15] There is also evidence that paternal smoking prior to conception is a risk factor for hepatoblastoma in children,[6,20] and there is more limited evidence for increased risk of acute lymphocytic leukaemia (ALL).[6,21]

The overall worldwide burden of disease from exposure to secondhand smoke has recently been estimated to have been 603 000 deaths in 2004, 47% of them in women, 28% in children and 26% in men, and accounting for about 1% of worldwide mortality.[22] A new report from the US Surgeon General on tobacco smoke (the 30th such report on the subject) provides a comprehensive review of the biology of diseases, through mechanisms of DNA damage, inflammation and oxidative stress, resulting from active and passive smoking.[23]

5.2 Hazardous Chemicals in Tobacco Smoke

There are several types of tobacco smoke: first, there is that which is inhaled directly by the smoker drawing on the cigarette (mainstream smoke); second, there is the stream of smoke released into the air while the smoker is not inhaling from the lit cigarette (sidestream smoke); and third, there is the smoke exhaled from the smoker's lungs (exhaled mainstream smoke). Secondhand smoke (or ETS) is thus composed of exhaled mainstream smoke and sidestream smoke, although the relative contribution of the two sources will vary depending on circumstances.

The chemical composition of tobacco smoke is extremely complex. Around 4000 compounds have been isolated from mainstream smoke.[5] These consist of those derived from the tobacco itself, from tobacco additives, the paper and filter and those formed as combustion products of them. At least 69 of them are genotoxic carcinogens, including a number that are classified by the IARC as Group 1 human carcinogens.[5,24] Examples of these are shown in Table 5.1.

In principle, many of the components of tobacco smoke could be used as biomarkers of active smoking and, potentially, of passive smoking. Structures of some of the more important ones, and some of their metabolites, are shown in Figure 5.1. Many of these are also present in other sources, so this needs to be taken into consideration when assessing their validity as biomarkers of active or passive smoking. An example of this is nicotine, the alkaloid that is the cause of tobacco addiction, which can also enter the body from dietary sources, including the consumption of aubergines and tomatoes.[5] However, non-tobacco sources of nicotine are estimated to make a negligible contribution to the total intake in smokers, and their contribution to the amount taken up by non-smokers exposed to secondhand smoke is unlikely to interfere with the use of nicotine or its major metabolite cotinine as biomarkers of passive smoking.[5]

Table 5.1 Carcinogens in tobacco smoke.

Class	*Examples*	*IARC Classification*[a]
Polycyclic aromatic hydrocarbons	Benzo[a]pyrene	1
	Naphthalene	2B
Heterocyclic hydrocarbons	Furan	2B
N-Nitrosamines	4-(Methylnitrosamino)-1-(3-pyridinyl)-1-butanone (NNK)	2B
	N-Nitrosonornicotine (NNN)	2B
Aromatic amines	2-Naphthylamine	1
	2-Toluidine	1
	4-Aminobiphenyl	1
N-Heterocyclic amines	2-Amino-1-methyl-6-phenylimidazo[4,5-*b*] pyridine (PhIP)	2B
Aldehydes	Formaldehyde	1
	Acetaldehyde	2B
Phenolic compounds	Catechol	2B
Volatile hydrocarbons	Benzene	1
	1,3-Butadiene	1
	Styrene	2B
	Isoprene	2B
Miscellaneous organic compounds	Acrylamide	2A
	Hydrazine	2B
	Urethane	2B
Metals and metal compounds	Arsenic	1
	Beryllium	1
	Nickel	1
	Chromium	1
	Cadmium	1
	Cobalt	2B
	Lead	2B
Radioisotope	Polonium-210	1[b]

[a] The IARC classification on the evaluation of carcinogenic risk to humans has the following definitions: Group 1, *carcinogenic to humans*; Group 2A, *probably carcinogenic to humans*; Group 2B, *possibly carcinogenic to humans*.
[b] Polonium-210 emits alpha particles, which are *carcinogenic to humans* (Group 1).

Cotinine in urine, saliva or blood is currently the most widely used biomarker of exposure to ETS.

Important carcinogenic components of tobacco smoke that do not derive from any other sources are the tobacco-specific nitrosamines. An example is 4-(methylnitrosamino)-1-(3-pyridinyl)-1-butanone (NNK), which undergoes metabolism to 4-(methylnitrosamino)-1-(3-pyridinyl)-1-butanol (NNAL) and its glucuronides (NNAL-Gluc). These compounds have formed the basis of assays to measure exposure to ETS, described below. Other carcinogenic compounds found in tobacco smoke, but also found in other sources, may be of some use for biomonitoring high exposure to tobacco smoke (*e.g.* from active smoking). These include the aromatic amines and polycyclic aromatic hydrocarbons (PAHs). However, as shall be described, their ability to distinguish

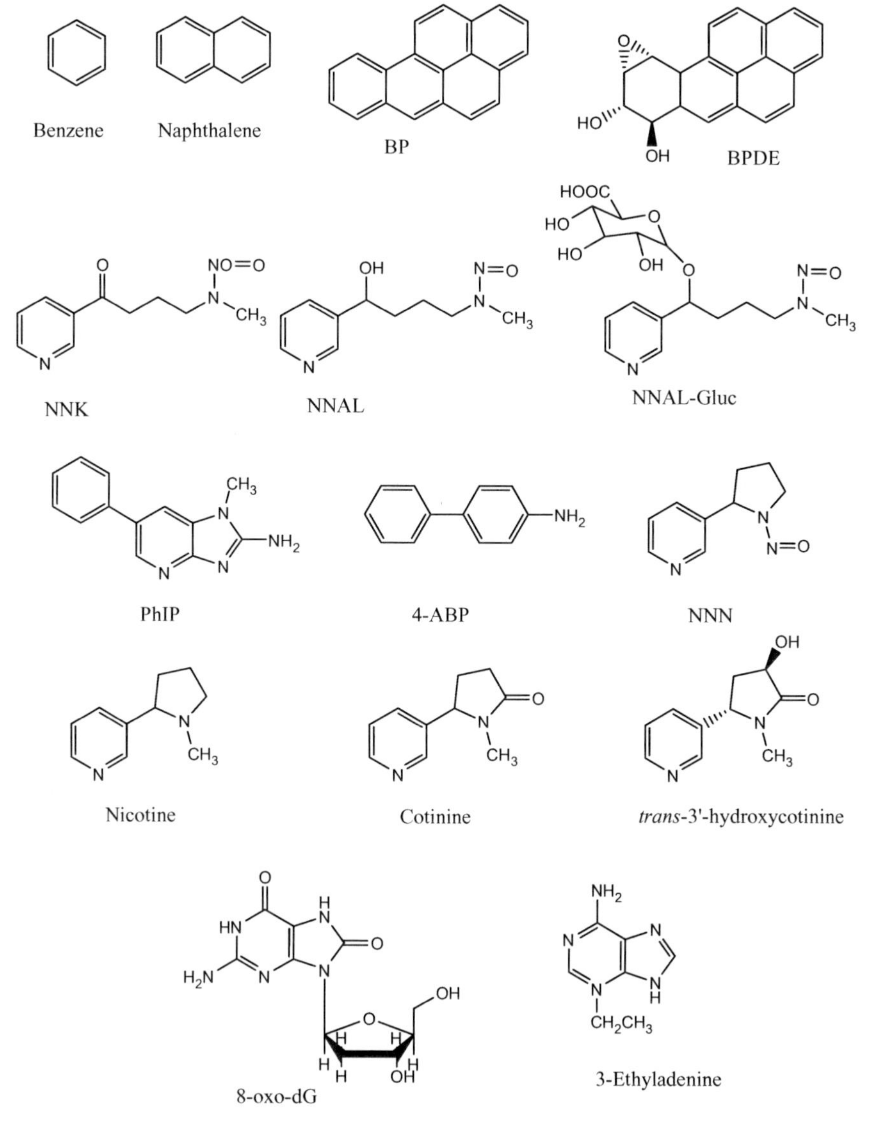

Figure 5.1 Components of tobacco smoke and some of their prominent metabolites with potential utility for biomonitoring human exposure to active and passive smoking.

exposed non-smokers from exposed ones has not been found adequate in some cases, making them less reliable biomarkers of exposure to ETS.

A newer concept that has emerged recently is that of *thirdhand smoke*. This derives from a process whereby secondhand smoke adsorbed on to surfaces (*e.g.* skin, furniture, clothing, automobile interiors) becomes desorbed subsequently, resulting in further human exposure. In addition, chemical reactions between atmospheric chemicals, including ozone, nitrous acid and nitric oxides, and the adsorbed tobacco smoke components can result in additional formation of hazardous chemicals, including carcinogenic tobacco-specific nitrosamines and other compounds not found in fresh tobacco smoke. For example, in a study exploring reactions between adsorbed nicotine and ambient nitrous acid, two nitrosamines found in tobacco smoke, 4-(methylnitrosamino)-1-(3-pyridinyl)-1-butanone (NNK) and *N*-nitrosonornicotine (NNN), were detected, as well as the novel component 1-(*N*-methyl-*N*-nitrosamino)-1-(3-pyridinyl)-4-butanal (NNA).[25]

5.3 Biomarkers of Involuntary Smoking

In order to consider the potential utility of a biomarker of smoking for monitoring exposure to secondhand smoke, it is first necessary to consider whether the biomarker can differentiate between active smokers and non-smokers. This can be asked in one of two ways: is the mean/median level observed in smokers significantly different from the mean level in non-smokers? Or, do the levels detected in individuals distinguish the smoker from the non-smoker? In the former case, there may be inter-individual variations such that there is overlap in values between the two groups, yet it may still be a useful biomarker in molecular epidemiological studies. However, only if there is no overlap between exposure groups can the biomarker be diagnostic of the smoking status of individuals. This needs to be borne in mind when distinguishing non-smokers exposed to secondhand smoke from those who are unexposed, where quantitative differences should logically be much less than between smokers and non-smokers.

In the following sections, substances and biological matrices that can be or have been used to investigate human exposure to secondhand tobacco smoke are described. In each case the strengths and weaknesses are mentioned, together with recent examples of their application. The levels at which the biomarkers have been detected are summarized in Table 5.2.

5.3.1 Urine

There are many studies in the literature in which cotinine or its parent compound nicotine have been measured in urine as a biomarker to distinguish active smokers, passive smokers and unexposed non-smokers (also cotinine in blood and saliva, as described later). It represents a relatively short-term measure of exposure because cotinine is excreted within 2–3 days of exposure.[26] Values are routinely corrected for urine volume by normalizing to the creatinine content of the urine; alternatively a 24-hour sample is taken and analysed. Cotinine levels are often higher in urine than in other matrices (*e.g.* blood and saliva) which may make it a more sensitive biomarker for exposure to ETS.[5] The relative ease with which samples can be obtained makes urine a particularly attractive matrix for monitoring exposure in children. The urinary half-life of

Table 5.2 Levels of tobacco chemicals and metabolites detected in human material; selected publications. CI, confidence intervals; COPD, chronic obstructive pulmonary disease; ETS, environmental tobacco smoke; GM, geometric mean; IQR, inter-quartile range; n.d., not detected; NNAL, 4-(methylnitrosamino)-1-(3-pyridinyl)-1-butanol; NNAL-Gluc, glucuronides of NNAL; NNN, *N'*-nitrosonornicotine; SD, standard deviation; SE, standard error.

Human material	*Compound*	*Levels and exposure*	*Study population*	*Reference*
Urine	Nicotine	Unexposed non-smokers 0.012 ± 0.010 (SD) nmol/mL ETS-exposed non-smokers 0.046 ± 0.062 (SD) nmol/mL	Non-smoking women	30
	Nicotine	Smokers 13.6 ± 8.22 (SD) nmol/mL	Adults	34
	Cotinine	10.93 (IQR 17.29) ng/mL	Children under 4 whose mothers smoked	27
	Cotinine	Unexposed non-smokers 0.009 ± 0.008 (SD) nmol/mL ETS-exposed non-smokers 0.045 ± 0.051 (SD) nmol/mL	Non-smoking women	30
	Cotinine	Passive smokers 1.82 ng/mg creatinine (IQR 0.45–7.33) Active smokers 1155 ng/mg creatinine (IQR 703–2715)	Adults from 3 countries	33
	Cotinine	Smokers 13.1 ± 6.71 (SD) nmol/mL	Adults	34
	Cotinine	No exposure 0.44 ng/mg creatinine (GM) (95% CI 0.24–0.80) Smoke exposure outside home 4.10 ng/mg creatinine (GM) (95% CI 2.33–7.23) Smoke exposure inside home 5.30 ng/mg creatinine (GM) (95% CI 3.25–8.63)	Children aged 3–27 months admitted to hospital with lower respiratory illness	67
	NNAL	0.601 ± 0.366 (SD) pmol/mL 0.857 ± 0.514 (SD) pmol/mg creatinine	Adults	34
	NNAL-Gluc	1.35 ± 0.738 (SD) pmol/mL 1.84 ± 0.879 (SD) pmol/mg creatinine	Adults	34
	NNAL + NNAL-Gluc	Unexposed non-smokers 0.007 ± 0.010 (SD) nmol/mL ETS-exposed non-smokers 0.050 ± 0.068 (SD) nmol/mL	Non-smoking women	30
	NNAL + NNAL-Gluc	Median 0.024 pmol/mg creatinine (range 0.0012–0.28)	Patients with COPD and self-reported exposure to ETS	26

Table 5.2 (*Continued*)

Human material	*Compound*	*Levels and exposure*	*Study population*	*Reference*
	NNAL + NNAL-Gluc	ETS-exposed non-smokers 5.56 pg/mL (95% CI 4.8–6.4) 5.27 pg/mg creatinine (95% CI 4.5–6.2)	Adults (NHANES cohort)	32
	NNAL + NNAL-Gluc	Passive smokers 5.19 pg/mg creatinine (IQR 2.04–11.6) Smokers 183 pg/mg creatinine (IQR 103–393)	Adults from 3 countries	33
	NNAL + NNAL-Gluc	0.24 ± 0.26 ng/mL	Smokers only	62
Plasma	Nicotine	9.56 ± 7.74 (SD) ng/mL	Smokers only	62
	Nicotine	0.72–1.03 ng/mL pre-exposure to ETS 1.09–1.51 ng/mL post-exposure to ETS	Non-smoking adults before and after visiting pubs	78
	Cotinine	<2ng/mL non-smokers reporting no or low exposure to ETS >2ng/mL non-smokers reporting low or high exposure to ETS	Adults	38
	Cotinine	Children of non-smoking mothers, no smoker in household 0.250 ± 0.124 (SE) ng/mL Children of non-smoking mothers, smoker in household 0.87 ± 0.203 (SE) ng/mL Children of mothers who smoke 4.14 ± 0.542 (SE) ng/mL Non-smoking mothers, no smoker in household 0.958 ± 0.790 (SE) ng/mL Non-smoking mothers, smoker in household 1.64 ± 0.967 (SE) ng/mL Smoking mothers 170 ± 21.2 (SE) ng/mL	Mothers and children	46
	Cotinine	No smoker in household 0.264 ± 0.596 (SD) ng/mL Smoker in household, non-smoking mother 0.869 ± 1.13 (SD) ng/mL Mother smokes 4.61 ± 7.19 (SD) ng/mL	Children (pre-school, aged 1–6 years)	47
	Cotinine	Mean values 1–4.5 ng/mL	Non-smoking adults in 3 different locations in Greece, with seasonal variation	57

	Cotinine	198 ± 134 (SD) ng/mL	Smokers only	62
	Cotinine	Non-smokers 0.609 ± 1.824 (SD) ng/mL Smokers 54.67 ± 29.63 (SD) ng/mL	Adults	64
	Cotinine	0–2.6 ng/mL pre-exposure to ETS (1/15 subjects +ve) 1.72–3.92 ng/mL post-exposure to ETS	Non-smoking adults before and after visiting pubs	78
	trans-3'-hydroxycotinine	69.5 ± 53.1 ng/mL	Smokers only	62
	Myosmine	Non-smokers 0.155 ± 0.183 (SD) ng/mL Smokers 0.296 ± 0.349 (SD) ng/ml	Adults	64
Serum	Cotinine	Non-smokers exposed to ETS, up to 14.0 ng/mL Smokers 138.4 ng/mL (mean)	Male adults	36
Saliva	Cotinine	Non-smokers 1.85 ± 4.50 (SD) ng/mL Smokers 97.43 ± 84.54 (SD) ng/ml	Adults	64
	Myosmine	Non-smokers 0.73 ± 0.65 (SD) ng/mL Smokers 2.54 ± 2.68 (SD) ng/mL	Adults	64
Toenails	Nicotine	No exposure 0.08 ng/mg (mean) Passive exposure 0.28 ng/mg (mean) Active smoking 1.71 ng/mg (mean) Active smoking plus passive exposure 2.18 ng/mg (mean)	Women (Nurses' Health Study)	58
	Nicotine	No exposure 0.10 ng/mg (median) Passive exposure 0.14 ng/mg (median) Active smoking 1.77 ng/mg (median)	Women (Nurses' Health Study)	59
	Nicotine	Non-smokers 0.09 ng/mg (mean) Smokers 5.9 ± 5.6 (SD) ng/mg	Adults	61
	Nicotine	5.44 ± 6.41 (SD) ng/mg	Smokers only	62
	Nicotine	Non-smokers 132 ± 82 (SD) ng/g Smokers 1971 ± 818 (SD) ng/g	Adults	64
	Cotinine	Non-smokers 0.01 ng/mg (mean) Smokers 1.6 ± 1.3 (SD) ng/mg	Adults	61
	Cotinine	0.67 ± 0.79 (SD) ng/mg	Smokers only	62
	Cotinine	Non-smokers n.d. Smokers 1237 ± 853 (SD) ng/g	Adults	64
	NNN	Non-smokers 0.35 ± 0.16 (SD) fmol/mg Smokers 4.63 ± 6.48 (SD) fmol/mg	Adults	63

Table 5.2 (*Continued*)

Human material	*Compound*	*Levels and exposure*	*Study population*	*Reference*
	NNAL	Non-smokers n.d. (<0.02 pg/mg) Smokers 0.41 ± 0.67 (SD) pg/mg	Adults	61
	NNAL	0.18 ± 0.22 pg/mg	Smokers only	62
	Myosmine	Non-smokers 21.2 ± 14.9 (SD) ng/g Smokers 65.9 ± 56.4 (SD) ng/g	Adults	64
Hair	Nicotine	Range 0.19–47.82 ng/mg No exposure 0.58 ng/mg (GM) (95% CI 0.49–0.68) Smoke exposure outside home 2.63 ng/mg (GM) (95% CI 2.03–3.40) Smoke exposure inside home 5.62 ng/mg (GM) (95% CI 4.60–6.85)	Children admitted to hospital with lower respiratory illnesses	67
	Nicotine	Range for non-smokers 0.06–1.82 ng/mg Range for smokers 0.91–63.5 ng/mg	Article is a review of 14 original studies	68
	Nicotine	Women 0.44 ng/mg (IQR 0.15–1.13) Children 0.80 ng/mg (IQR 0.27–2.24)	Women and children living in households with smokers	69
	Nicotine	Non-exposed non-smoking mothers 0.32 ng/mg (mean) ETS-exposed non-smoking mothers 0.99 ng/mg (mean) Smoking mothers 9.61 ng/mg (mean) Infants of non-exposed non-smoking mothers ~0.4 ng/mg Infants of ETS-exposed non-smoking mothers ~0.4 ng/mg Infants of smoking mothers ~1.0 ng/mg	Mother–baby pairs	72
Meconium	Nicotine	Infants of unexposed mothers 1.74 ng/g (95% CI 1.50–2.01) Infants of ETS-exposed mothers 2.56 ng/g (95% CI 1.67–3.94) Infants of smoking mothers 26.8 ng/g (95% CI 2.44–7.73)	Newborn infants	74

	Cotinine	Infants of unexposed mothers 0.10 ng/g (95% CI 0.08–0.12) Infants of ETS-exposed mothers 0.20 ng/g (95% CI 0.09–0.40) Infants of smoking mothers 15.9 ng/g (95% CI 7.0–36.2)	Newborn infants	74
	trans-3'-hydroxycotinine	Infants of unexposed mothers 0.08 ng/g (95% CI 0.06–0.11) Infants of ETS-exposed mothers 0.15 ng/g (95% CI 0.07–0.34) Infants of smoking mothers 23.3 ng/g (95% CI 9.5–56.8)	Newborn infants	74
Teeth	Nicotine	Children of non-smoking parents 15.0 ng/g (mean) Children with one smoking parent 24.6 ng/g (mean) Children with two smoking parents 43.5 ng/g (mean)	Children aged 7.4 ± 0.6 yrs	76
	Nicotine	Children of non-smoking parents 14.11 ng/g (mean) Children with one smoking parent 22.51 ng/g (mean) Children with two smoking parents 27.49 ng/g (mean)	Children	77
	Cotinine	Children of non-smoking parents 13.9 ng/g (mean) Children with one smoking parent 18.6 ng/g (mean) Children with two smoking parents 28.6 ng/g (mean)	Children aged 7.4 ± 0.6 yrs	76
	Cotinine	Children of non-smoking parents 8.76 ng/g (mean) Children with one smoking parent 11.25 ng/g (mean) Children with two smoking parents 11.57 ng/g (mean)	Children	77

nicotine in newborns is 3–4 times longer in adults, but the half-life of cotinine is about the same.[5]

To test whether counselling mothers would be an effective strategy for reducing children's exposure to ETS, a double-blind controlled trial was conducted among 108 smoking mothers who exposed their children to tobacco smoke in the home.[27] Both the counselled group and the control group reported reducing their children's exposure, but at 12 months the reported exposure in the counselled group was 41.2% of that reported by the control group. However, children's urine cotinine levels reduced only slightly in the counselled group and increased nearly two-fold in the control group. Thus it could be concluded that even if counselling did not actually decrease exposure of children (as monitored by urinary cotinine levels), it did at least prevent an increase in exposure in the study population.

The detection in human urine of tobacco carcinogens and their metabolites has been comprehensively reviewed by Hecht.[28] Several of these have utility in distinguishing smokers from non-smokers, including *trans,trans*-muconic acid and *S*-phenylmercapturic acid (metabolites of benzene), naphthols (metabolites of naphthalene), 1-hydroxypyrene, thioethers (formed from a variety of conjugated electrophiles), 3-ethyladenine (which indicates the presence in tobacco smoke of an unidentified ethylating agent[29]), NNAL and NNAL-Gluc (metabolites of tobacco-specific nitrosamines). An advantage of monitoring exposure to carcinogenic components of tobacco smoke in urine, rather than the non-carcinogen cotinine, is that it can provide insight into carcinogen uptake and metabolism, and the potential influence of genetic polymorphisms in modifying risk.[28] However, most of the metabolites listed above have only limited utility for monitoring exposure to ETS, because they can come from other sources besides tobacco smoke, and these other sources can mask the possible tobacco smoke exposure of non-smokers. The exceptions are NNAL and NNAL-Gluc; these are metabolites of the tobacco-specific nitrosamine NNK. Mass spectrometry assays for them are both sensitive and specific.[28] The levels of the two compounds in the urine of non-smokers who live with smokers were found to be significantly higher than the levels in the urine of non-smokers in households where no one smoked.[30] The levels in the passive smokers were, on average, 5.6% of those of the smoking partners,[30] which is reasonably consistent with the estimated excess risk of lung cancer from passive smoking being about 2% of that of smokers.[31]

Although levels of NNAL plus NNAL-Gluc in the urine of non-smokers correlate well with cotinine levels in urine[28] and serum,[32] it is important to note for passive smoking that the ratio of NNAL to cotinine is much higher compared with that for active smokers.[33] This suggests that if cotinine is used as a surrogate for exposure of non-smokers to carcinogens (or, at least, to one of them, NNK), then exposure will be underestimated. Urinary NNAL is reported to be a longer-term measure of ETS exposure than urinary cotinine,[26] the former having an elimination half-life of 40–45 days.[34]

It is well established that the urine of smokers has mutagenic activity in bacteria, and the urine of non-smokers exposed to ETS has been found to be mutagenic in some, but not all, studies investigating its mutagenicity.[5] As expected, increases in mutagenic activity were small relative to those in smokers, and paralleled urinary cotinine levels. Other, confounding, exposures (*e.g.* dietary or occupational) could also give rise to mutagenic activity of urine in some individuals. In view of this, there does not seem to be much advantage in using bacterial mutagenicity as a biomarker of exposure in preference to the measurement of cotinine in urine, although it does demonstrate that passive smoking involves exposure to genotoxicants, which emphasizes the fact that ETS is hazardous to non-smokers.

5.3.2 Blood

5.3.2.1 Compounds and Metabolites

Plasma (or serum) cotinine is a widely used marker of exposure, albeit short term, to secondhand smoke.[5] Although levels are generally lower in plasma than in urine, there is a good correlation between the two.[35]

In a prospective population based study[36] of the risk of coronary heart disease, compared with serum cotinine concentrations in men who said they did not smoke, relative hazards for heart disease were 1.45 [95% confidence interval (CI) 1.01–2.08], 1.49 (1.03–2.14) and 1.57 (1.08–2.28) for the second, third and fourth quartiles of cotinine concentrations, respectively, relative to the first quartile. Thus higher concentrations of serum cotinine among non-smokers are associated with an excess risk of coronary heart disease of around 50–60%, and the risk associated with passive smoking appears to be widespread in this population (7735 men aged 40–59 years in 24 towns in England, Scotland and Wales). There was, however, no consistent association between cotinine and risk of stroke in this study.

Levels of carboxyhaemoglobin in erythrocytes, and of thiocyanates in serum, distinguish smokers from non-smokers, but they do not reliably distinguish non-smokers exposed to ETS from those unexposed; there are other environmental sources of both these compounds, which confound their validity as biomarkers of exposure to ETS.[5]

5.3.2.2 Protein Adducts

Many of the genotoxic chemical carcinogens in tobacco smoke undergo metabolic activation to electrophilic species, through which they exert their deleterious biological effects. For example, NNK is metabolically activated to reactive species via intermediate formation of NNAL, and the PAH benzo[a]pyrene (BP) is metabolized to its ultimately reactive species benzo[a]pyrene diol-epoxide (BPDE) (shown in Figure 5.1). A result of the formation of these reactive species in cells is the covalent modification of cellular macromolecules, including protein and DNA. These chemically modified entities are termed adducts. If they are chemically stable, they can serve as medium-term (*i.e.*

weeks to months) biomarkers of exposure to tobacco smoke. When smokers and non-smokers have been compared for their levels of 4-aminobiphenyl (4-ABP)-haemoglobin adducts, measured by gas chromatography–mass spectrometry (GC-MS), not only were the levels higher in the smokers, but also there was no overlap in values with those found in non-smokers.[37] However, in some small studies in which non-smokers with self-reported exposure to ETS were compared with unexposed non-smokers, there was not a significant difference in 4-ABP-haemoglobin adduct levels except when the exposed subjects had detectable cotinine levels.[38,39] Thus while it is clear that tobacco smoke is the major source of 4-ABP-haemoglobin adducts in smokers, the sources that are the major contributors to the presence of the adduct in non-smokers are less clear.[40] Diet and traffic exhaust have been proposed as sources, in addition to ETS,[40] and a recent study concluded that ETS is not a major source of 4-ABP-haemoglobin adducts in non-smokers.[41]

Nevertheless, some studies have reported that 4-ABP–haemoglobin adducts can reflect exposure to ETS. Measurements of maternal–fetal exchange of 4-ABP in smoking ($n = 14$) and non-smoking ($n = 38$) pregnant women found that levels of the haemoglobin adduct in cord blood were consistently lower than in maternal blood. A significant correlation was found by linear regression between maternal and fetal levels; this was true for all subjects ($p < 0.001$) and for smokers only ($p = 0.002$) but not for non-smokers only ($p = 0.06$).[42] A further report on this same sample explored the relationship between adduct levels and ETS exposure of the mothers. The level of exposure was determined by questionnaire, diary and personal air monitor and reported as <0.5, 0.5–1.9 and ≥ 2.0 $\mu g/m^3$ weekly average nicotine concentration. Among 40 non-smoking women there was a significant correlation between 4-ABP–haemoglobin and exposure category ($p = 0.009$).[43] In two other reports, lower levels of adducts were also found in fetal blood, relative to maternal blood, at levels that correlated with the smoking status of the mothers.[44,45]

In another study, exposure of young (pre-school) children to ETS from their mothers' smoking was investigated by measuring plasma cotinine levels and PAH–albumin adducts in peripheral blood, the latter detected by ELISA. The study involved 87 mother–child pairs; 31 mothers smoked and 56 did not. Not only did the mothers who smoked have higher levels of adducts but the levels in the children of the smoking mothers were also significantly higher ($p < 0.05$). There was also a significant correlation between the adduct levels in the mothers and the levels in their children ($p = 0.014$).[46] In a subsequent study, PAH–albumin adducts were measured by ELISA and 4-ABP–haemoglobin adducts were measured by GC-MS. In 109 pre-school children, PAH–albumin adducts were significantly higher in the children whose mothers smoked ($p = 0.001$) or who lived with other smokers ($p = 0.017$), compared with those in non-smoking households; levels of 4-ABP–haemoglobin adducts were also higher in these children, but not significantly ($p = 0.073$ and 0.066, respectively).[47]

In an investigation of the formation of other protein adducts, blood samples from smoking and non-smoking mothers and cord blood from their newborns were analysed for HOEtVal in haemoglobin.[48] Adduct levels were significantly higher in the smoking women than in the non-smokers, and the concentrations in the newborns were significantly higher ($p < 0.01$) in those with smoking mothers [$n = 13$, 147 ± 68 (mean ± SD) pmol/g] than in those with non-smoking mothers ($n = 10$, 42 ± 18 pmol/g). There was also a significant correlation ($p < 0.01$) between adduct levels in the infants and the levels in their mothers. The same maternal–fetal samples were subsequently analysed for CEVal formed by acrylonitrile.[49] Adduct levels were significantly higher in the smoking mothers than in the non-smoking ones and there was a significant correlation between the adduct levels in the smoking mothers and those in their newborns ($p < 0.001$). The levels were approximately two-fold lower in the babies than in their smoking mothers and there was a significant correlation between the levels in the babies and the numbers of cigarettes smoked per day by the mothers ($p = 0.009$). For the babies of non-smoking mothers, CEVal levels were below the limit of detection of the assay.

In a study of 69 adults, smokers comprised 27 of the group and of the 42 non-smokers, 19 were classified as passive smokers by self-reporting and cotinine levels. Although the levels of BPDE adducts with albumin and haemoglobin were higher in the smokers (differences in the former were statistically significant, but in the latter were not) there was no difference found between non-smokers exposed to ETS and those not or rarely exposed.[50]

Levels of HOEtVal in haemoglobin, while higher in smokers than in non-smokers, were found to be present at the same levels in non-smokers who did not live or work with a smoker ($n = 74$) as in those who did ($n = 28$).[51]

Overall, these studies present a mixed picture of the ability of protein adducts to detect passive exposure to tobacco smoke. Thus their utility as biomarkers of exposure to ETS is limited.

5.3.2.3 DNA Adducts

In many of the tissues of smokers, the mean levels of DNA adducts are elevated relative to those in non-smokers;[37] this reflects the systemic distribution of tobacco smoke components and the fact that smoking causes cancer in many organs.[5,6] In such studies, there is considerable inter-individual variation in adduct levels in both smokers and non-smokers, and considerable overlap in values between the two groups. In contrast to the number of studies that have investigated the levels of DNA adducts, or other measures of DNA damage, in the tissues of smokers, ex-smokers and non-smokers, relatively few studies have employed these biomarkers to monitor exposure of non-smokers to ETS. Considering that the magnitude of the difference between smokers and non-smokers is often quite small (for example, less than two-fold in the case of DNA adducts in white blood cells), it is probable that in many circumstances this biomarker may lack the sensitivity to distinguish the effects of exposure to ETS from exposure to other sources of environmental carcinogens. Studies in which a distinction has been made between non-smokers exposed to ETS and those unexposed are described below.

In a study in which declining DNA adduct levels in the white blood cells of smokers enrolled in a smoking cessation programme were measured by enzyme-linked immunosorbent assay (ELISA) using antibodies to BPDE–DNA adducts, levels of adducts both at baseline and 10 weeks after cessation were significantly associated with hours of exposure to ETS at home and were also higher if the subject lived with another smoker ($p = 0.009$, 0.02 and 0.02, respectively). However, there was no observable influence of exposure to ETS in the workplace.[52]

Using the prevalence of serum antibodies to BPDE–DNA adducts in an Italian population as a biomarker of exposure to environmental PAHs, investigators reported an association between % positivity and smoking status (25.7% of smokers, 21.5% of ex-smokers and 17.8% of non-smokers) and the living environment (26.0% urban dwellers and 17.9% suburban dwellers), but no association with passive smoking was found.[53]

In a study in which significant differences were observed between the levels of adducts, determined by a flow cytometric method using BPDE–DNA antibodies, in peripheral lymphocytes of smokers ($n = 40$) and non-smokers ($n = 35$), the mean value for non-smokers with no or low exposure to ETS ($n = 17$) was lower than that of non-smokers who were exposed ($n = 18$), but the difference was not statistically significant.[54]

A group of five non-smokers was exposed to ETS under controlled conditions (exposure to gas phase only for 8 hours, followed by exposure to whole ETS for 8 hours, 40 hours later). When their monocyte DNA was analysed by ^{32}P-postlabelling, no changes in the adduct patterns were seen after either exposure period, compared with the samples obtained before exposure.[55] Similarly, the frequency of DNA single-strand breaks in mononuclear cells of non-smokers was not affected by controlled exposure to ETS for 8 hours.[56]

In a study of biomarkers of air pollution exposure in two Greek populations,[57] one from an urban location, the other rural, ^{32}P-postlabelling analysis of lymphocyte DNA revealed the presence of bulky PAH-like DNA adducts that were more abundant in the rural subjects than in the urban ones. The variation in adduct levels was found to significantly ($p < 0.001$) parallel self-reported exposure to ETS and plasma cotinine concentrations rather than exposure to other environmental sources of PAHs.

Overall, as with protein adducts, these studies do not suggest that DNA adducts levels are sufficiently different between unexposed and ETS-exposed smokers to make them reliable biomarkers of ETS exposure. While target tissues for carcinogenicity might be more suitable than blood cells as a source of DNA,[37] the opportunities to obtain such tissues from healthy individuals are very limited.

5.3.3 Saliva

Numerous studies have found that cotinine levels in the saliva of non-smokers exposed to ETS are significantly higher than in that from non-exposed non-smokers.[5] This is a short-term measure of exposure, giving information on exposure in the preceding hours. Heavily exposed workers, for example bar workers, may have peak levels more characteristic of smokers.[5] For monitoring exposure in children, sampling saliva has the clear advantage of being less invasive than taking blood samples, although obtaining sufficient material from very young children may be a problem.

5.3.4 Breath

Analysis of breath can provide a non-invasive, but short-term, means of monitoring secondhand exposure to tobacco smoke. Thus, concentrations of carbon monoxide (CO) exhaled by non-smokers who have spent the day in a smoking environment are significantly higher than levels exhaled by unexposed non-smokers, as are the levels in women exposed during pregnancy and their newborn infants.[5] On the other hand, exhaled concentrations of nitric oxide (NO) are *lower* after exposure to secondhand tobacco smoke, both for adults and for newborns exposed prenatally.[5] Exhaled benzene does not appear to be a reliable indicator of exposure to secondhand tobacco smoke, at least in the home.[5]

5.3.5 Toenails and Fingernails

Uptake of tobacco components by nails offers another relatively long-term biomarker of smoking exposure, and obtaining nail clippings is less invasive than obtaining blood samples. Toenails are less likely to suffer from surface contamination than fingernails, or hair (see Section 5.3.6). Toenail clippings from women participating in the Nurses' Health Study have been analysed for their nicotine content in a number of studies. After washing with dichloromethane, the samples are digested overnight in 1M NaOH and analysed by high-performance liquid chromatography (HPLC). In a pilot study, toenails from 106 women were analysed, and a significant correlation with self-reported exposure to tobacco smoke (active smoking, passive smoking, no exposure) was found.[58] A subsequent larger study included samples from 2485 subjects.[59] The study found that the median level of nicotine in toenails of non-smokers without exposure to secondhand smoke was 0.10 ng/mg, for non-smokers with passive exposure it was 0.14 ng/mg, while for smokers the median level was 1.77 ng/mg. However, there was considerable overlap in levels between the groups, with 5th–95th percentile ranges of 0.05–0.41, 0.05–0.83 and 0.24–6.34 ng/mg, respectively. In a nested case–control study of coronary heart disease also carried out in the Nurses' Health Study cohort,[60] involving 905 cases with two matched controls for each, toenail nicotine levels were significantly associated with risk of disease. The authors of these studies[58–60] have concluded that the analysis of nicotine levels in toenails has utility for future epidemiological studies, in particular where smoking history is not available or could be subject to biased reporting.

Other investigators,[61,62] using GC-MS, showed that cotinine and 4-(methylnitrosamino)-1-(3-pyridyl)-1-butanol (NNAL) are also detectable in human toenails. The levels of these, and of nicotine, in toenails correlated with their levels in plasma. It is not yet clear, however, what the utility of this method may be for monitoring exposure to secondhand smoke. The same investigators[63] have also developed a method for detecting *N'*-nitrosonornicotine (NNN) in toenails. The ratio of levels of NNN to NNAL was 2.8, distinctly different from the ratio found in analyses of urine (0.1). Whether this biomarker has utility for monitoring secondhand smoke exposure has yet to be determined.

Myosmine, another tobacco alkaloid that can undergo nitrosation *in vivo*, is detectable in toenails, and also in plasma and saliva.[64] Although the mean level in smokers is significantly higher than in non-smokers, the fold increase is less than for nicotine, and there was an overlap in the values of the two groups, which weakens the usefulness of myosmine as a biomarker of exposure to secondhand smoke.

Nails collected from newborn children of mothers who admitted to smoking during pregnancy have been found to contain nicotine and cotinine.[65] In the same study, a number of children of self-reported non-smoking mothers also had detectable levels of one or both compounds present in their nails; however, it was not clear from the study whether this was due to passive smoking by the mothers during pregnancy, or failure to admit to active smoking.

The fingernails of heavy smokers are often characteristically stained yellow–brown from direct exposure to tobacco smoke, and smoking cessation leads to the so-called harlequin nail, with its distinct line of demarcation between pigmented nail and the new growth of unstained nail.[66] For toenails, however, the uptake of nicotine and other tobacco constituents is likely to be principally via the blood.[61] Thus analysis of toenails for one or more of nicotine, cotinine, NNAL and NNN may be a useful means of monitoring exposure to secondhand smoke, but further validation studies are required.

5.3.6 Hair

Scalp hair grows at a rate of about 1 cm per month. It can thus provide a relatively long-term biomarker of exposure to toxicants, of the order of months, rather than hours or days in the case of urine, blood or saliva. Hair may be subject to adsorption from the atmosphere of tobacco smoke components, but this can generally be dealt with by prior extraction with dichloromethane. Although cotinine is present in hair, it is at lower concentrations than nicotine, which has thus been proposed as an alternative to measuring urinary cotinine.[67] There have been many studies on hair as a biomarker for exposure to tobacco smoke (reviewed by Al-Delaimy[68]). Some extreme forms of hair treatment, such as bleaching, may affect nicotine content, but in general, hair has good potential for monitoring exposure to secondhand smoke in adults and children. In an international study of non-smoking women and children exposed to secondhand smoke at home, involving subjects in 31 countries, it was estimated that there was an increase of 1% in women and 3% in children of hair nicotine for every 1 g/m^3 increase in air nicotine concentration.[69] The hair of newborn infants may be a source for biomonitoring prenatal exposure spanning the third trimester, although it may not always be obtainable in sufficient quantity.[70]

In a study of 210 mothers and their babies,[71] nicotine analysis of the mothers' hair distinguished between smokers, passive smokers and non-exposed non-smokers, and urine cotinine levels correlated with self-reported smoking status. Nicotine levels in the hair of the infants of women who had smoked during pregnancy were significantly higher than in infants of non-smokers, but did not distinguish maternal passive smoking during pregnancy from the non-exposed mothers. Analysis of preterm birth and neonatal outcomes in this same study population revealed several effects of prenatal exposure to ETS: infants had a lower birth weight and there was a greater risk of preterm birth, immediate newborn complications and respiratory distress syndrome (RDS).[72]

5.3.7 Fingerprints

Nicotine is detectable in smokers' fingerprints by GC-MS.[73] Smokers can be reliably distinguished from non-smokers by this technique, provided the fingerprints are relatively fresh (up to 1 hour old); the short half-life of nicotine in fingerprints ($\sim$11 hours) means that the prints of smokers and non-smokers are harder to distinguish by about 24 hours. Nevertheless the method has been used to demonstrate that nicotine can be transferred from smokers to non-smokers by indirect (*e.g.* via door handles) and direct (*e.g.* handshaking)[73] contact. However, its utility as a biomarker of passive smoking would appear to be limited.

5.3.8 Meconium

There is considerable interest in meconium, the first faeces of newborn babies, as a biological matrix for detecting prenatal exposure to a wide variety of toxicants. It provides information on exposures during the second and third trimesters of pregnancy, and its collection in the first 5 days after birth is easy and non-invasive. In a study in which mothers self-reported their exposure to tobacco smoke (active, passive, or unexposed) during pregnancy, the levels of nicotine, cotinine and *trans*-3'-hydroxycotinine were measured in the infants' meconium.[74] Levels of the compounds were 1–3 orders of magnitude higher in the meconium of infants born to mothers who smoked during pregnancy, and slightly higher where the mothers reported exposure to secondhand smoke, compared with those who were unexposed. These levels were positively associated with serum cotinine levels in the mothers and, interestingly, there was an inverse correlation between meconium concentrations and birth weight.[74] Thus the concentrations of tobacco metabolites in meconium appear to provide a useful measure of prenatal exposure to tobacco smoke.

5.3.9 Other Tissues and Matrices

In addition to those tissues and other human material already described, a number of other biological matrices have been proposed for the particular purpose of monitoring childhood and prenatal exposure to ETS. These include amniotic fluid, cord blood, breast milk and deciduous (children's) teeth.[70]

Cord blood can be used to measure cotinine levels, but will only give an indication of exposure to tobacco smoke in the last few days of pregnancy, although it can distinguish between active smoking, passive smoking and non-exposure during this short time period (the same is true of neonatal urine, although the results are less consistent than for cord blood).[70] Cotinine measurements of amniotic fluid can also distinguish smoking status of mothers and demonstrate fetal exposure, but collecting this specimen is an invasive procedure, unless performed at the time of birth, and thus not a suitable matrix for widespread biomonitoring.[70]

Placenta is another potential source of DNA for monitoring active and passive smoking relating to prenatal exposure. However, when placental DNA was analysed both for bulky DNA adducts by ^{32}P-postlabelling and for 8-oxo-dG (a DNA lesion indicative of oxidative damage to DNA) by electrochemical detection, neither method showed a difference between 11 smokers, 10 non-smokers and 9 non-smokers exposed to passive smoking.[75]

Among children of smoking mothers, breast feeding is a more important source of exposure to nicotine than passive smoking (by inhalation). Cotinine levels in breast milk have been linked to a number of ill-health effects in breast-fed infants.[70] Nevertheless, monitoring levels in breast milk can be used to encourage mothers who smoke to refrain from doing so shortly before feeding, in order to reduce exposure of the nursing infant to tobacco toxicants and carcinogens.[70]

Children's teeth could provide a biomarker of cumulative exposure to ETS from fetal life up to 6–8 years. Analysis of nicotine (by GC-MS) in children's teeth has indeed demonstrated such exposure in the children of smoking parents (identified by questionnaire).[76] Cotinine was a less sensitive biomarker in these circumstances, a result confirmed in a more recent study using a more sensitive tandem MS assay for both nicotine and cotinine.[77]

There are also a few studies reported in the literature in which other tissues and biological matrices have been investigated for monitoring exposure to ETS in adults. In a study of a small group of non-smokers, who provided blood and sputum samples before, and 1–2 hours after, spending 3 hours exposed to ETS, their plasma cotinine and nicotine levels were both significantly increased by the exposure, but the levels of DNA adducts in their peripheral blood lymphocytes were not.[78] There were, however, some changes observed in the adduct profiles of sputum DNA of some individuals, including the formation of an adduct characteristic of the major BPDE–DNA adduct in 3/15 individuals.[78]

With smoking now recognized as a causative factor for cancer of the ovary,[6] it is interesting to note that when immunohistochemical staining using antibodies to BPDE–DNA adducts was used on ovarian granulosa-lutein cells obtained from women undergoing *in vitro* fertilization (IVF) treatment, staining was intermediate in intensity for non-smoking women exposed to passive smoking ($n = 7$) between the values for smokers ($n = 14$) and those of unexposed non-smokers ($n = 11$), and was statistically significantly different from both these groups ($p < 0.0001$).[79]

5.4 Concluding Remarks

In the last 50 years it has become abundantly clear that tobacco smoking is not only harmful to those who smoke, but also to non-smokers who are exposed to tobacco smoke. This awareness has created a need for accurate and sensitive means of monitoring such exposure among non-smokers. Some countries have enacted legislation to limit the exposure of non-smokers to ETS; many others have yet to do so.

Although many biomarkers have been developed and validated as measures of tobacco smoke exposure in smokers, the number of studies in which these have been applied to monitoring exposure to secondhand smoke is relatively small. Thus there is scope for more research to refine and develop these assays for this more demanding application.

Cotinine measurement in blood, saliva or urine is currently the most widely used biomarker for exposure to ETS, but it provides information only on very recent exposure. There is thus the potential for long-termer biomarkers to be applied more widely, and the analysis of toenails and hair, for example, needs to be exploited more fully. Given that many of the carcinogenic chemicals in tobacco smoke can also be found in other sources, focusing on those that are not, for example, the tobacco-specific nitrosamines, may be an approach that offers greater specificity.

In view of the harm that prenatal and early-life exposure to tobacco smoke can cause, it is particularly important that objective measures of the efficacy of strategies to eliminate such exposures be developed further and applied. There are a number of tissues and biological matrices that can be utilized for this.

Currently there is no biomarker that can measure long-term exposure, over years or decades, to ETS. In the realms of worker compensation, for example for workers exposed to ETS in the hospitality industry, retrospective exposure assessment is likely to have to rely, for the foreseeable future, on approximations of past exposure calculated by mathematical modelling.[80]

However for future studies, particularly for prospective studies, biomarkers will be integral to their design. The ease of sample collection, transport and storage will be matters to consider when setting up such studies. The stability of the biomarkers during long-term storage will also be a factor to consider.

Acknowledgements

Research in the author's laboratory is funded principally by Cancer Research UK. Stan Venitt is thanked for critical reading of the manuscript.

References

1. N. L. Benowitz, Nicotine addiction, *N. Engl. J. Med.*, 2010, **362**, 2295–2303.
2. R. Doll, R. Peto, J. Boreham and I. Sutherland, Mortality in relation to smoking: 50 years' observations on male British doctors, *BMJ*, 2004, **328**, 1519.

3. R. Doll and A. B. Hill, The mortality of doctors in relation to their smoking habits; a preliminary report, *Br. Med. J.*, 1954, **1**, 1451–1455.
4. IARC, *IARC Monographs on the Evaluation of the Carcinogenic Risk of Chemicals to Humans, vol. 38, Tobacco Smoking*, International Agency for Research on Cancer, Lyon, 1986.
5. IARC, *IARC Monographs on the Evaluation of the Carcinogenic Risk of Chemicals to Humans, vol. 83, Tobacco Smoke and Involuntary Smoking*, International Agency for Research on Cancer, Lyon, 2004.
6. B. Secretan, K. Straif, R. Baan, Y. Grosse, F. El Ghissassi, V. Bouvard, L. Benbrahim-Tallaa, N. Guha, C. Freeman, L. Galichet and V. Cogliano, A review of human carcinogens – Part E: tobacco, areca nut, alcohol, coal smoke, and salted fish, *Lancet Oncol.*, 2009, **10**, 1033–1034.
7. J. He, S. Vupputuri, K. Allen, M. R. Prerost, J. Hughes and P. K. Whelton, Passive smoking and the risk of coronary heart disease – a meta-analysis of epidemiologic studies, *N. Engl. J. Med.*, 1999, **340**, 920–926.
8. T. F. Pechacek and S. Babb, How acute and reversible are the cardiovascular risks of secondhand smoke? *BMJ*, 2004, **328**, 980–983.
9. CDC, Annual smoking-attributable mortality, years of potential life lost, and economic costs – United States, 1995-1999. *MMWR Morb. Mortal. Wkly Rep.*, 2002, **51**, 300–303.
10. R. Bonita, J. Duncan, T. Truelsen, R. T. Jackson and R. Beaglehole, Passive smoking as well as active smoking increases the risk of acute stroke, *Tob. Control*, 1999, **8**, 156–160.
11. D. F. Mackay, M. O. Irfan, S. Haw and J. P. Pell, Meta-analysis of the effect of comprehensive smoke-free legislation on acute coronary events, *Heart*, 2010, **96**, 1525–1530.
12. C. M. Rudin, E. Avila-Tang and J. M. Samet, Lung cancer in never smokers: a call to action, *Clin. Cancer Res.*, 2009, **15**, 5622–5625.
13. J. M. Samet, E. Avila-Tang, P. Boffetta, L. M. Hannan, S. Olivo-Marston, M. J. Thun and C. M. Rudin, Lung cancer in never smokers: clinical epidemiology and environmental risk factors, *Clin. Cancer Res.*, 2009, **15**, 5626–5645.
14. H. C. Kinney and B. T. Thach, The sudden infant death syndrome, *N. Engl. J. Med.*, 2009, **361**, 795–805.
15. C. E. Hunt and F. R. Hauck, Sudden infant death syndrome, *CMAJ*, 2006, **174**, 1861–1869.
16. J. B. Daly, J. H. Wiggers, S. Burrows and M. Freund, Household smoking behaviours and exposure to environmental tobacco smoke among infants: are current strategies effectively protecting our young? *Aust. N. Z. J. Public Health*, 2010, **34**, 269–273.
17. J. R. DiFranza, C. A. Aligne and M. Weitzman, Prenatal and postnatal environmental tobacco smoke exposure and children's health, *Pediatrics*, 2004, **113**, 1007–1015.
18. J. M. Rogers, Tobacco and pregnancy, *Reprod. Toxicol.*, 2009, **28**, 152–160.

19. M. Ekblad, M. Gissler, L. Lehtonen and J. Korkeila, Prenatal smoking exposure and the risk of psychiatric morbidity into young adulthood, *Arch. Gen. Psychiatry*, 2010, **67**, 841–849.
20. D. Pang, R. McNally and J. M. Birch, Parental smoking and childhood cancer: results from the United Kingdom Childhood Cancer Study, *Br. J. Cancer*, 2003, **88**, 373–381.
21. K. M. Lee, M. H. Ward, S. Han, H. S. Ahn, H. J. Kang, H. S. Choi, H. Y. Shin, H. H. Koo, J. J. Seo, J. E. Choi, Y. O. Ahn and D. Kang, Paternal smoking, genetic polymorphisms in CYP1A1 and childhood leukemia risk. *Leuk. Res.*, 2009, **33**, 250–258.
22. M. Oberg, M. S. Jaakkola, A. Woodward, A. Peruga and A. Pruss-Ustun, Worldwide burden of disease from exposure to second-hand smoke: a retrospective analysis of data from 192 countries. *Lancet*, 2010, epub ahead of print Nov 26.
23. US Department of Health and Human Services, *How Tobacco Smoke Causes Disease: The Biology and Behavioral Basis for Smoking-Attributable Disease: A Report of the Surgeon General*, Atlanta, GA, 2010.
24. R. Baan, Y. Grosse, K. Straif, B. Secretan, F. El Ghissassi, V. Bouvard, L. Benbrahim-Tallaa, N. Guha, C. Freeman, L. Galichet and V. Cogliano, A review of human carcinogens – Part F: chemical agents and related occupations, *Lancet Oncol.*, 2009, **10**, 1143–1144.
25. M. Sleiman, L. A. Gundel, J. F. Pankow, P. Jacob, 3rd, B. C. Singer and H. Destaillats, Formation of carcinogens indoors by surface-mediated reactions of nicotine with nitrous acid, leading to potential thirdhand smoke hazards, *Proc. Natl. Acad. Sci. USA*, 2010, **107**, 6576–6581.
26. M. D. Eisner, P. Jacob, 3rd, N. L. Benowitz, J. Balmes and P. D. Blanc, Longer term exposure to secondhand smoke and health outcomes in COPD: impact of urine 4-(methylnitrosamino)-1-(3-pyridyl)-1-butanol, *Nicotine Tob. Res.*, 2009, **11**, 945–953.
27. M. F. Hovell, J. M. Zakarian, G. E. Matt, C. R. Hofstetter, J. T. Bernert and J. Pirkle, Effect of counselling mothers on their children's exposure to environmental tobacco smoke: randomised controlled trial, *BMJ*, 2000, **321**, 337–342.
28. S. S. Hecht, Human urinary carcinogen metabolites: biomarkers for investigating tobacco and cancer, *Carcinogenesis*, 2002, **23**, 907–922.
29. V. Prevost and D. E. Shuker, Cigarette smoking and urinary 3-alkyladenine excretion in man, *Chem. Res. Toxicol.*, 1996, **9**, 439–444.
30. K. E. Anderson, S. G. Carmella, M. Ye, R. L. Bliss, C. Le, L. Murphy and S. S. Hecht, Metabolites of a tobacco-specific lung carcinogen in non-smoking women exposed to environmental tobacco smoke, *J. Natl. Cancer Inst.*, 2001, **93**, 378–381.
31. W. J. Blot and J. K. McLaughlin, Passive smoking and lung cancer risk: what is the story now? *J. Natl. Cancer Inst.*, 1998, **90**, 1416–1417.
32. J. T. Bernert, J. L. Pirkle, Y. Xia, R. B. Jain, D. L. Ashley and E. J. Sampson, Urine concentrations of a tobacco-specific nitrosamine

carcinogen in the U.S. Population from secondhand smoke exposure, *Cancer Epidemiol. Biomarkers Prev.*, 2010, 2969–2977.

33. N. L. Benowitz, M. Goniewicz, M. Eisner, E. Lazcano-Ponce, W. Zielinska-Danch, B. Koszowski, A. Sobczak, C. Havel and P. Jacob, 3rd, Urine cotinine underestimates exposure to tobacco-derived lung carcinogen 4-(methylnitrosamino)-1-(3-pyridyl)-1-butanone (NNK) in passive compared to active smokers, *Cancer Epidemiol. Biomarkers Prev.*, 2010, 2795–2800.
34. S. S. Hecht, S. G. Carmella, M. Chen, J. F. Dor Koch, A. T. Miller, S. E. Murphy, J. A. Jensen, C. L. Zimmerman and D. K. Hatsukami, Quantitation of urinary metabolites of a tobacco-specific lung carcinogen after smoking cessation, *Cancer Res.*, 1999, **59**, 590–596.
35. N. L. Benowitz, K. M. Dains, D. Dempsey, B. Herrera, L. Yu and P. Jacob, 3rd, Urine nicotine metabolite concentrations in relation to plasma cotinine during low-level nicotine exposure, *Nicotine Tob. Res.*, 2009, **11**, 954–960.
36. P. H. Whincup, J. A. Gilg, J. R. Emberson, M. J. Jarvis, C. Feyerabend, A. Bryant, M. Walker and D. G. Cook, Passive smoking and risk of coronary heart disease and stroke: prospective study with cotinine measurement, *BMJ*, 2004, **329**, 200–205.
37. D. H. Phillips, Smoking-related DNA and protein adducts in human tissues, *Carcinogenesis*, 2002, **23**, 1979–2004.
38. M. Maclure, R. B. Katz, M. S. Bryant, P. L. Skipper and S. R. Tannenbaum, Elevated blood levels of carcinogens in passive smokers, *Am. J. Public Health*, 1989, **79**, 1381–1384.
39. H. Bartsch, N. Caporaso, M. Coda, F. Kadlubar, C. Malaveille, P. Skipper, G. Talaska, S. R. Tannenbaum and P. Vineis, Carcinogen hemoglobin adducts, urinary mutagenicity, and metabolic phenotype in active and passive cigarette smokers, *J. Natl. Cancer Inst.*, 1990, **82**, 1826–1831.
40. E. Richter and B. Branner, Biomonitoring of exposure to aromatic amines: haemoglobin adducts in humans, *J. Chromatogr. B Analyt. Technol. Biomed. Life Sci.*, 2002, **778**, 49–62.
41. M. K. Schorp and D. E. Leyden, Biomonitoring of smoke constituents: exposure to 4-aminobiphenyl and 4-aminobiphenyl hemoglobin adduct levels in nonsmokers and smokers, *Inhal. Toxicol.*, 2010, **22**, 725–737.
42. J. Coghlin, P. H. Gann, K. Hammond, P. L. Skipper, K. Taghizadeh, M. Paul and S. R. Tannenbaum, 4-Aminobiphenyl hemoglobin adducts in fetuses exposed to the tobacco smoke carcinogen in utero, *J. Natl. Cancer Inst.*, 1991, **83**, 274–280.
43. S. K. Hammond, J. Coghlin, P. H. Gann, M. Paul, K. Taghizadeh, P. L. Skipper and S. R. Tannenbaum, Relationship between environmental tobacco smoke exposure and carcinogen-hemoglobin adduct levels in nonsmokers, *J. Nat. Cancer Inst.*, 1993, **85**, 474–478.
44. S. R. Myers, J. A. Spinnato, M. T. Pinorini-Godly, C. Cook, B. Boles and G. C. Rodgers, Characterization of 4-aminobiphenyl-hemoglobin adducts in maternal and fetal blood-samples, *J. Toxicol. Environ. Health*, 1996, **47**, 553–566.

45. M. T. Pinorini-Godly and S. R. Myers, HPLC and GC/MS determination of 4-aminobiphenyl haemoglobin adducts in fetuses exposed to the tobacco smoke carcinogen in utero, *Toxicology*, 1996, **107**, 209–217.
46. F. G. Crawford, J. Mayer, R. M. Santella, T. B. Cooper, R. Ottman, W. Y. Tsai, G. Simon-Cereijido, M. Wang, D. Tang and F. P. Perera, Biomarkers of environmental tobacco smoke in preschool children and their mothers, *J. Natl. Cancer Inst.*, 1994, **86**, 1398–1402.
47. D. Tang, D. Warburton, S. R. Tannenbaum, P. Skipper, R. M. Santella, G. S. Cereijido, F. G. Crawford and F. P. Perera, Molecular and genetic damage from environmental tobacco smoke in young children, *Cancer Epidemiol. Biomarkers Prev.*, 1999, **8**, 427–431.
48. R. Tavares, P. Ramos, J. Palminha, M. A. Bispo, I. Paz, A. Bras, J. Rueff, P. B. Farmer and E. Bailey, Transplacental exposure to genotoxins. Evaluation in haemoglobin of hydroxyethylvaline adduct levels in smoking and non-smoking mothers and their newborns, *Carcinogenesis*, 1994, **15**, 1271–1274.
49. R. Tavares, H. Borba, M. Monteiro, M. J. Proenca, N. Lynce, J. Rueff, E. Bailey, G. M. Sweetman, R. M. Lawrence and P. B. Farmer, Monitoring of exposure to acrylonitrile by determination of N-(2-cyanoethyl)valine at the N-terminal position of haemoglobin, *Carcinogenesis*, 1996, **17**, 2655–2660.
50. G. Scherer, S. Frank, K. Riedel, I. Meger-Kossien and T. Renner, Biomonitoring of exposure to polycyclic aromatic hydrocarbons of non-occupationally exposed persons, *Cancer Epidemiol. Biomarkers Prev.*, 2000, **9**, 373–380.
51. R. Bono, M. Vincenti, V. Meineri, C. Pignata, U. Saglia, O. Giachino and E. Scursatone, Formation of N-(2-hydroxyethyl)valine due to exposure to ethylene oxide via tobacco smoke: A risk factor for onset of cancer, *Environ. Res.*, 1999, **81**, 62–71.
52. L. A. Mooney, R. M. Santella, L. Covey, A. M. Jeffrey, W. Bigbee, M. C. Randall, T. B. Cooper, R. Ottman, W. Y. Tsai and L. Wazneh, Decline of DNA damage and other biomarkers in peripheral blood following smoking cessation, *Cancer Epidemiol. Biomarkers Prev.*, 1995, **4**, 627–634.
53. S. Petruzzelli, A. Celi, N. Pulera, F. Baliva, G. Viegi, L. Carrozzi, G. Ciacchini, M. Bottai, F. Di Pede, P. Paoletti and C. Giuntini, Serum antibodies to benzo(a)pyrene diol epoxide-DNA adducts in the general population: effects of air pollution, tobacco smoking, and family history of lung diseases, *Cancer Res.*, 1998, **58**, 4122–4126.
54. R. Shinozaki, S. Inoue, K. S. Choi and T. Tatsuno, Association of benzo[a]pyrene-diol-epoxide-deoxyribonucleic acid (BPDE-DNA) adduct level with aging in male smokers and nonsmokers, *Arch. Environ. Health*, 1999, **54**, 79–85.
55. O. Holz, T. Krause, G. Scherer, U. Schmidt-Preuss and H. W. Rudiger, 32P-postlabelling analysis of DNA adducts in monocytes of smokers and passive smokers, *Int. Arch. Occup. Environ. Health*, 1990, **62**, 299–303.
56. O. Holz, R. Meissner, M. Einhaus, F. Koops, K. Warncke, G. Scherer, F. Adlkofer, E. Baumgartner and H. W. Rudiger, Detection of DNA

single-strand breaks in lymphocytes of smokers, *Int. Arch. Occup. Environ. Health*, 1993, **65**, 83–88.
57. P. Georgiadis, J. Topinka, M. Stoikidou, S. Kaila, M. Gioka, K. Katsouyanni, R. Sram, H. Autrup and S. A. Kyrtopoulos, Biomarkers of genotoxicity of air pollution (the AULIS project): bulky DNA adducts in subjects with moderate to low exposures to airborne polycyclic aromatic hydrocarbons and their relationship to environmental tobacco smoke and other parameters, *Carcinogenesis*, 2001, **22**, 1447–1457.
58. W. K. Al-Delaimy, G. N. Mahoney, F. E. Speizer and W. C. Willett, Toenail nicotine levels as a biomarker of tobacco smoke exposure, *Cancer Epidemiol. Biomarkers Prev.*, 2002, **11**, 1400–1404.
59. W. K. Al-Delaimy and W. C. Willett, Measurement of tobacco smoke exposure: comparison of toenail nicotine biomarkers and self-reports, *Cancer Epidemiol. Biomarkers Prev.*, 2008, **17**, 1255–1261.
60. W. K. Al-Delaimy, M. J. Stampfer, J. E. Manson and W. C. Willett, Toenail nicotine levels as predictors of coronary heart disease among women, *Am. J. Epidemiol.*, 2008, **167**, 1342–1348.
61. I. Stepanov, R. Feuer, J. Jensen, D. Hatsukami and S. S. Hecht, Mass spectrometric quantitation of nicotine, cotinine, and 4-(methylnitrosamino)-1-(3-pyridyl)-1-butanol in human toenails, *Cancer Epidemiol. Biomarkers Prev.*, 2006, **15**, 2378–2383.
62. I. Stepanov, S. S. Hecht, B. Lindgren, P. Jacob, 3rd, M. Wilson and N. L. Benowitz, Relationship of human toenail nicotine, cotinine, and 4-(methylnitrosamino)-1-(3-pyridyl)-1-butanol to levels of these biomarkers in plasma and urine, *Cancer Epidemiol. Biomarkers Prev.*, 2007, **16**, 1382–1386.
63. I. Stepanov and S. S. Hecht, Detection and quantitation of N'-nitrosonornicotine in human toenails by liquid chromatography-electrospray ionization-tandem mass spectrometry, *Cancer Epidemiol. Biomarkers Prev.*, 2008, **17**, 945–948.
64. K. Schutte-Borkovec, C. W. Heppel, A. K. Heling and E. Richter, Analysis of myosmine, cotinine and nicotine in human toenail, plasma and saliva, *Biomarkers*, 2009, **14**, 278–284.
65. F. Mari, L. Politi and E. Bertol, Nails of newborns in monitoring drug exposure during pregnancy, *Forensic Sci. Int.*, 2008, **179**, 176–180.
66. A. Verghese, G. Krish, D. Howe and M. Stonecipher, The harlequin nail. A marker for smoking cessation, *Chest*, 1990, **97**, 236–238.
67. W. K. Al-Delaimy, J. Crane and A. Woodward, Is the hair nicotine level a more accurate biomarker of environmental tobacco smoke exposure than urine cotinine? *J. Epidemiol. Community Health*, 2002, **56**, 66–71.
68. W. K. Al-Delaimy, Hair as a biomarker for exposure to tobacco smoke, *Tob. Control*, 2002, **11**, 176–182.
69. S. Kim, H. Wipfli, A. Navas-Acien, F. Dominici, E. Avila-Tang, G. Onicescu, P. Breysse and J. M. Samet, Determinants of hair nicotine concentrations in nonsmoking women and children: a multicountry study of secondhand smoke exposure in homes, *Cancer Epidemiol. Biomarkers Prev.*, 2009, **18**, 3407–3414.

70. H. Llaquet, S. Pichini, X. Joya, E. Papaseit, O. Vall, J. Klein and O. Garcia-Algar, Biological matrices for the evaluation of exposure to environmental tobacco smoke during prenatal life and childhood, *Anal. Bioanal. Chem.*, 2010, **396**, 379–399.
71. K. B. Ashford, E. Hahn, L. Hall, M. K. Rayens, M. Noland and R. Collins, Measuring prenatal secondhand smoke exposure in mother-baby couplets, *Nicotine Tob. Res.*, 2010, **12**, 127–135.
72. K. B. Ashford, E. Hahn, L. Hall, M. K. Rayens, M. Noland and J. E. Ferguson, The effects of prenatal secondhand smoke exposure on preterm birth and neonatal outcomes, *J. Obstet. Gynecol. Neonatal Nurs.*, 2010, **39**, 525–535.
73. M. Benton, M. J. Chua, F. Gu, F. Rowell and J. Ma, Environmental nicotine contamination in latent fingermarks from smoker contacts and passive smoking, *Forensic Sci. Int.*, 2010, **200**, 28–34.
74. J. M. Braun, J. L. Daniels, C. Poole, A. F. Olshan, R. Hornung, J. T. Bernert, Y. Xia, C. Bearer, D. B. Barr and B. P. Lanphear, A prospective cohort study of biomarkers of prenatal tobacco smoke exposure: the correlation between serum and meconium and their association with infant birth weight, *Environ. Health*, 2010, **9**, 53.
75. H. Daube, G. Scherer, K. Riedel, T. Ruppert, A. R. Tricker, P. Rosenbaum and F. Adlkofer, DNA adducts in human placenta in relation to tobacco smoke exposure and plasma antioxidant status, *J. Cancer Res. Clin. Oncol.*, 1997, **123**, 141–151.
76. O. Garcia-Algar, O. Vall, J. Segura, J. A. Pascual, D. Diaz, L. Mutnoz, P. Zuccaro, R. Pacifici and S. Pichini, Nicotine concentrations in deciduous teeth and cumulative exposure to tobacco smoke during childhood, *JAMA*, 2003, **290**, 196–197.
77. E. Marchei, X. Joya, O. Garcia-Algar, O. Vall, R. Pacifici and S. Pichini, Ultrasensitive detection of nicotine and cotinine in teeth by high-performance liquid chromatography/tandem mass spectrometry, *Rapid Commun. Mass Spectrom.*, 2008, **22**, 2609–2612.
78. A. Besaratinia, L. M. Maas, E. M. Brouwer, E. J. Moonen, T. M. De Kok, G. J. Wesseling, S. Loft, J. C. Kleinjans and F. J. Van Schooten, A molecular dosimetry approach to assess human exposure to environmental tobacco smoke in pubs, *Carcinogenesis*, 2002, **23**, 1171–1176.
79. M. T. Zenzes, L. A. Puy and R. Bielecki, Immunodetection of benzo[a]pyrene adducts in ovarian cells of women exposed to cigarette smoke, *Mol. Hum. Reprod.*, 1998, **4**, 159–165.
80. S. Kolb, U. Bruckner, D. Nowak and K. Radon, Quantification of ETS exposure in hospitality workers who have never smoked, *Environ. Health*, 2010, **9**, 49.

CHAPTER 6

Biomarkers of Exposure: Mycotoxins – Aflatoxin, Deoxynivalenol and Fumonisins

PAUL C TURNER,*[,a] LIANA VAN DER WESTHUIZEN[b] AND ANDRE NOGUEIRA DA COSTA[c]

[a] School of Public Health, University of Maryland, College Park, MD, USA; [b] PROMEC Unit, Medical Research Council, P.O. Box 19070, Tygerberg, South Africa; [c] Molecular Carcinogenesis Group, Section of Mechanisms of Carcinogenesis, International Agency for Research on Cancer Lyon, France

6.1 Background

Fungal colonization of dietary staples is common in all regions of the world and causes a significant financial burden where crops are not suitable for market. The types of fungal colonization are numerous and diverse and their presence reflects in part the type of crop, the crop stress, the climate, and the local harvesting and the storage conditions.[1,2] Once fungal contamination has occurred, specific temperature and humidity conditions promote the production of highly toxic secondary metabolites (mycotoxins); and in fact approximately 25% of the world's cereal crops are predicted to be contaminated.[1,2] There are many families of mycotoxins, but those of particular concern to animal welfare and human health include the aflatoxins, deoxynivalenol and the fumonisins. Owing to the frequency and levels of contamination, demonstrated toxicity, and the differences in the stage of biomarker development for exposure to mycotoxins, these three were chosen as the focus of this chapter. Aflatoxin exposure biomarkers were validated approximately 20 years ago, and

Issues in Toxicology No. 10
Biomarkers and Human Biomonitoring, Volume 2: Selected Biomarkers of Current Interest
Edited by Lisbeth E. Knudsen and Domenico Franco Merlo

Published by the Royal Society of Chemistry, www.rsc.org

the focus here includes their use in descriptive epidemiology; a urinary deoxynivalenol biomarker was recently validated and the focus here is on descriptive biomonitoring; while candidates for fumonisin biomarkers are considered in detail. Readers are also encouraged to examine the literature on exposure biomarkers to ochratoxin A, an important nephrotoxic *Aspergillus* mycotoxin.[3] The intake of ochratoxin A has been correlated with urinary levels but not with serum levels of ochratoxin A,[4] thus the former should be used for biomonitoring. Ochratoxin A is not considered further in this review, and readers should be aware that there are many additional important mycotoxins[1,2] for which no biomarkers are reported in the literature.

6.2 Aflatoxins

6.2.1 Aflatoxin Structure

The aflatoxins are a family of highly substituted coumarins containing a fused dihydrofurofuran moiety. Aflatoxin B1 (AFB1) is the most frequently occurring and the most toxic and carcinogenic; whilst other members of the family include AFB2, AFG1, and AFG2, see Figure 6.1. A number of *Aspergillus* strains produce aflatoxins in hot and humid climates though *A. flavus* (producing AFB1 and AFB2) and *A. parasiticus* (producing AFB1, AFB2, AFG1 and AFG2) dominate the natural production[5]. *A. flavus* occurs throughout the world, whilst *A. parasiticus* is restricted mainly to Africa, South America, Central America and North America. Small amounts of aflatoxins are prepared for research purposes, though no other legal uses occur. The production of aflatoxins in biological warfare as a terror weapon has also been reported.[6]

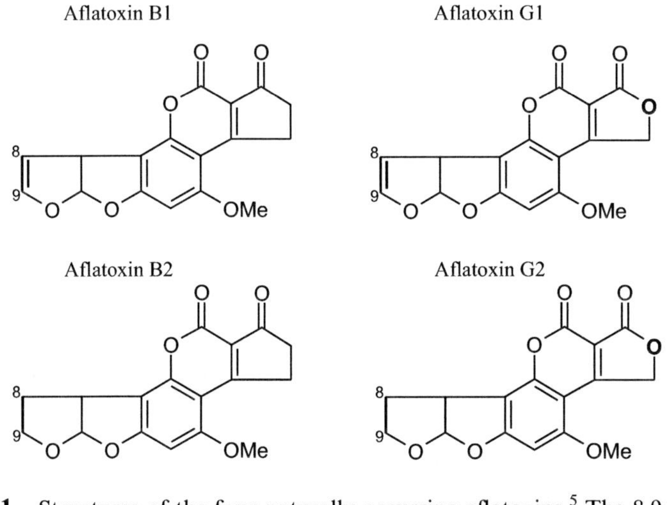

Figure 6.1 Structures of the four naturally occurring aflatoxins.[5] The 8,9 position is where a reactive epoxide can be readily formed across the double bond.

6.2.2 Aflatoxins and Diet

While numerous foods items can be contaminated with aflatoxins, maize and groundnuts (peanuts) are the mostly frequently contaminated food items, often at high levels in many developing countries; these items are also the most important dietary staples in many of the poorest regions at risk of producing contaminated food. Aflatoxins are stable in food preparations and exposure occurs most frequently through consumption of these contaminated dietary staples.[5,7] Aflatoxin production occurs both in the field during crop growth, and during storage. Prolonged storage (months) in poor conditions, with hot and humid environments, favours both *Aspergillus* infestation and aflatoxin production; thus rural peasants reliant on subsistence farming in more tropical world regions, with limited storage capabilities (simple huts), are at particular risk. Such populations frequently have more restricted diets than many in wealthier regions, thus critical components of exposure in West Africa, for example, are:

- frequent mycotoxin contamination of stored food;
- high levels of contamination;
- frequent consumption of contaminated food – their dietary staples;
- limited availability of non-contaminated alternatives.

In addition, in circumstances where peasant farmers produce a moderate excess of crops such that they are able to sell part of their harvest, the healthier looking crops are most frequently sold, whilst the poor quality may be retained in family storage.[7] More moderate levels of exposure occur in world regions with intermediate wealth, such as Egypt and Brazil,[8–12] which is in part a reflection of the availability of greater dietary variety, but also better harvesting and storage capabilities. Exposure in developed countries is relatively low and infrequent,[13,14] partly owing to dietary variety, but also because of the financial resources to regulate and restrict any contaminated crops getting into the food chain. Aflatoxin metabolism gives rise to a variety of metabolites, see Figures. 6.2 and 6.3, including aflatoxin M1, which is the predominant metabolite in

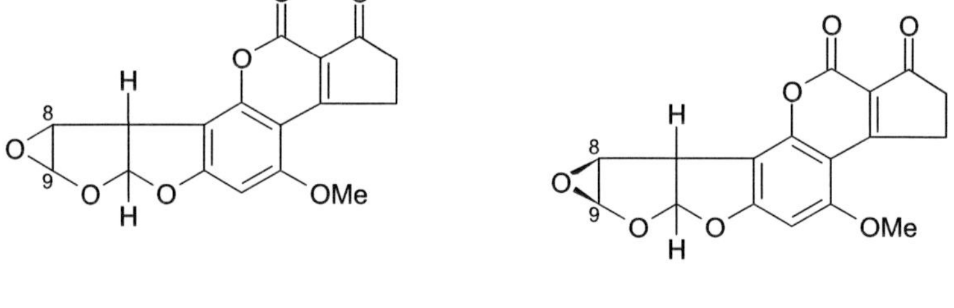

Aflatoxin B1 8,9 exo-epoxide (toxic and mutagenic)

Aflatoxin B1 8,9 endo-epoxide (toxic)

Figure 6.2 Structure of the aflatoxin B1 epoxides.

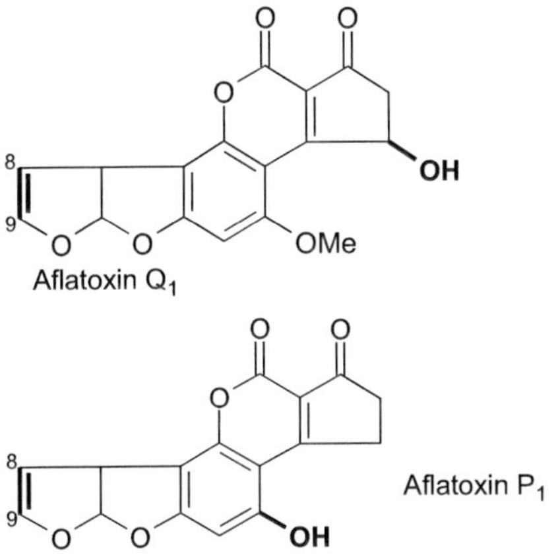

Figure 6.3 Structures of additional aflatoxin metabolites from phase 1 metabolism. Measurement of aflatoxin M1 in urine, but not Q1 or P1, provides a validated exposure biomarker for aflatoxin B1 ingestion.

milk of exposed lactating animals, including human breast milk.[5] Thus in regions with highly contaminated food and feed, both animal milk and breast milk provide additional potential routes of exposure.

6.2.3 Aflatoxin Metabolism and Biomarkers

Aflatoxin B1 is the most frequently occurring of the aflatoxins, and the most toxic, and thus its metabolism has been studied most thoroughly and is best understood.[5,7,15] Aflatoxins B1 and G1 undergo a number of possible transformations, although the toxicity and carcinogenicity is best understood in terms of activation to an epoxide at the double bond at the 8,9 position, see Figures. 6.1 and 6.2. The epoxide is highly reactive and causes damage to cells and macromolecules by covalently binding to proteins and nucleic acids. Activation occurs through cytochrome P450 (CYP) 1A2, 3A4, 3A5 (adult forms) and 3A7 (the fetal form of 3A4) and generates two reactive species: an exo-8,9-epoxide and an endo-8,9-epoxide. The CYP1A2 metabolism of AFB1 generates the hydroxylated metabolite, aflatoxin M1 (AFM1), in addition to a mixture of the AFB1 8,9-epoxides, predominantly the AFB1 endo-8,9-epoxide.[16] CYP3A4 generates some AFB1 exo-8,9-epoxide, though the predominant metabolite is the less toxic hydroxylation product aflatoxin Q1 (AFQ1).[17] CYP3A5 predominantly forms the carcinogenic exo-8,9-epoxide, and a lesser proportion of AFQ1. CYP3A7 is the fetal form of CYP3A4[18] and is observed in the fetus within approximately 2 months of conception.[19] CYP3A7 generally declines within the first postnatal year,[19] and CYP3A4/5 reach approximately 30% of adult levels about 1 month after birth; although CYP3A7 has occasionally been detected in adult liver.[20] Thus at an early *in utero* stage, throughout pregnancy and into infancy, any aflatoxin that is transferred to the fetus (or ingested by the young infant) can be metabolized to the toxic and carcinogenic endo- and exo-epoxides. It is thought that aflatoxin B2 and G2 will in part be reduced to aflatoxins B1 and G1, prior to activation to form reactive epoxides.[21]

As a result of its lack of mutagenic potential, the importance of the endo-epoxide appears somewhat neglected in the literature by cancer-driven molecular epidemiologists. Whilst it is not able to covalently modify DNA, it will cause severe toxicity, necrosis and subsequent increased cell proliferation, particularly in the intestinal tract and liver. The exo-epoxide causes similar toxicity, but it is also mutagenic, forming a stable covalent adduct with the N7 moiety of guanine.[22] Depurination at this site releases 8,9-dihydro-8-(N7guanyl)-9-hydroxy aflatoxin B1 (AFB1-N7-Gua), see Figure 6.4, which is observed in the urine of individuals exposed naturally to aflatoxin through their diet.[23,24] While DNA adduct formation is clearly important in terms of carcinogenicity, the relative contribution of generalized non-specific organ toxicity in the aetiology of aflatoxin-driven cancer remains poorly examined.

Aflatoxin exo- and endo-epoxides are detoxified through conjugation with glutathione *via* a family of glutathione S-transferases,[25] and are eventually excreted as the mercapturic acid. Non-enzymic hydrolysis of both epoxides to aflatoxin B1-8,9-dihydrodiol also occurs, although a putative enzymic role for epoxide hydrolase in this step remains controversial. The aflatoxin B1-8,9-dihydrodiol can ring open to form a dialdehyde which is toxic, but not mutagenic. The dialdehyde can be reduced *via* aflatoxin aldehyde reductase (AFAR) to the less toxic dialchohol.[26,27] Polymorphisms in aflatoxin metabolizing enzymes and changes in expression by other environmental influences, including diet, cause inter-individual variations in aflatoxin metabolism.[21,28] Aflatoxin dihydrodiol can undergo a slow base-catalysed ring opening to result in a resonating dialdehyde phenolate ion, which is capable of forming adducts with protein amino groups, particularly lysine (Figure 6.5),[29] and one such protein adduct, known as aflatoxin–albumin, is observed in the sera of individuals exposed to aflatoxin.[11–14,29–46]

Aflatoxin-N7-guanine in urine, aflatoxin–albumin in sera and aflatoxin M1 have been demonstrated to reflect the level of aflatoxin B1 intake, and these measures represent validated exposure biomarkers for biomonitoring.[23,24,30–33,47]

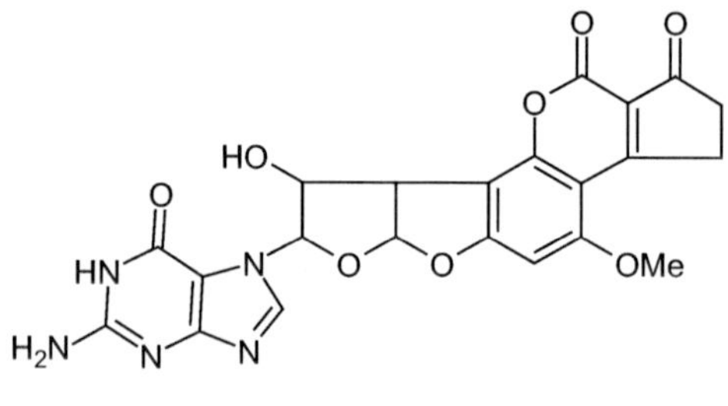

Figure 6.4 Structure of aflatoxin-N7-guanine formed by the aflatoxin exo-epoxide. binding to guanine in DNA.

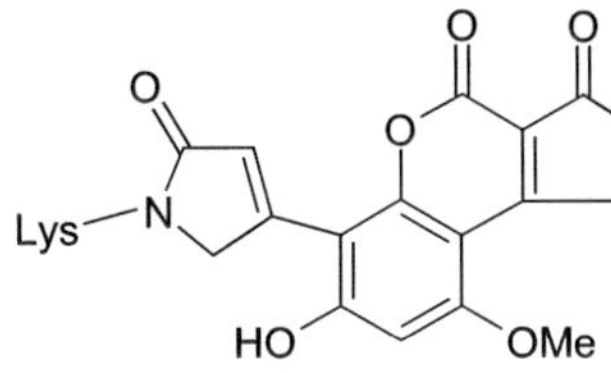

Figure 6.5 Structure of the major pronase digest product (aflatoxin–lysine) from the aflatoxin–albumin adduct.

In studies measuring the natural exposure to the toxin through diet in high risk settings about 1–3% of the aflatoxin ingested is converted to the aflatoxin–albumin adduct,[31,32] 1–2% is found as urinary aflatoxin M1,[48] 0.1–0.4% as breast-milk aflatoxin M1,[49] and 0.2% as aflatoxin-N7-guanine.[32] The aflatoxin M1 breast-milk survey was limited to only five samples and it should not be regarded as a validated biomarker, although it is a bio-measure that is of concern with regard to the feeding infant. Aflatoxins B1, B2, G1, G2, Q1, P1, aflatoxin-mercapturic acid and aflatoxicol have all been detected in urine, and the above species and aflatoxin M1 have been detected in breast milk,[5] indicative of aflatoxin exposure, although no dose–response relationship with aflatoxin has been demonstrated; thus none of these latter bio-measures represents a validated exposure biomarker.

6.2.4 Geographical Variations in Aflatoxin Biomarkers

Owing to the half-life of aflatoxin–albumin being ~20 days, it represents an integrated measure of exposure over a period of several months.[5,13,29–31] It has also been shown to be stable in cryopreserved samples over a period of 25 years,[33] a critical attribute for any biomarker that is utilized with bio-bank repositories. It also the only mycotoxin bio-measure reported for which a set of samples have undergone analysis in multiple laboratory settings, including the use of different analytical techniques.[37,38] Excellent correlations between laboratory analyses were reported in both studies. It is probably the biomarker that has been used to biomonitor aflatoxin exposure in the most diverse geographical settings, including high, moderate and low risk countries, and a variety of age groups. Table 6.1 highlights some of these data to demonstrate the variety of settings, exposure frequencies and exposure levels that are encountered.[11–14,36–46] The table is not exhaustive and predominantly reports data that used the same methodology for analysis. In general aflatoxin–albumin adducts are not detected in developed regions of the world,[13,14,36] are detected at moderate levels and frequency in countries such as Thailand,[13] Brazil[12] and Egypt,[11] and are very frequently detected (and at high levels) in some parts of Asia and Africa.[13,37–46] The frequency of detection and levels of urinary aflatoxin-N7-guanine and urinary aflatoxin M1, where measured, typically reflect those of aflatoxin–albumin.[47]

Table 6.1 Aflatoxin–albumin (AF–alb) adduct level and frequency (% Pos) in selected geographical regions.

Continent	*n*	*AF–alb (pg/mg) Mean (range)*	*% Pos*	*Age group*	*Reference*
Europe North America					
France/Poland	74	0	0	Mixed	13
Canada	200	0	0	Adults	14
USA (Texas)	184	<1 (nd–17)	21	Adults	36
South America					
Brazil	50	15[a] (nd–57)	62	Adult	12
Africa					
Benin and Togo	479	33 (nd–1064)	99	Child	37
Benin[c]	200	37 (nd–688)	98	Child	38
	200	39 (nd–744)	99		
	200	88 (5–1568)	100		
Ghana	785	11 (nd–269)	>75	Adult	39
Guinea	124	9 (nd–262)	96	Child	40
Egypt	46	6[a] (nd–33)	67	Adult	11
Kenya	155	43 (nd–580)	65	Mixed	13
Senegal					
The Gambia	119	40 (5–261)	100	Preg	41
	99	10 (5–190)	49	Cord	
	118	9 (5–30)	11	16wk	
The Gambia	128	60 (nd–391)	92	52wk	42
The Gambia	466	24 (nd–456)	93	Child	43
The Gambia	444	41 (3–459)	100	Child	44
The Gambia	391	57 (nd–720)	83	Child	45
Asia					
China-Quangxi	143	39 (nd–437)	69	Adult	13
-Shangdong	69	0	0		
Nepal	46	9 (nd–18)	15	Adult	13
Taiwan[d]	264	22 (5–356)[b]	ns	Adult	46
	264	14 (5–205)	ns	Adult	
Thailand	160	12 (nd–50)	11	Adults	13

[a] Mean of positives only.
[b] Units are fmol/mg.
[c] 200 Children measured at three time points within an 8-month period.
[d] 264 Adults measured at two time points separated by on average more than 1.5 years.

6.2.5 Aflatoxin and Liver Cancer

Aflatoxins were first discovered about 50 years ago because of a high frequency of liver toxicity and subsequent mortality of turkeys, termed Turkey X disease, in southern England. The causative agent was eventually traced to a naturally occurring feed contaminant within groundnuts (peanuts),[50,51] which was identified as a novel fungal toxin from *Aspergillus flavus* and was named aflatoxin.[52] At about the same time the investigation of a high incidence of liver cancer in farmed trout in the USA revealed the same toxin, albeit from contaminated cottonseed rather than groundnuts.[53] Turkey X disease affected

>100 000 animals and highlighted the importance of this novel naturally occurring dietary hazard. At that time the role of aflatoxins in human health was not established, and following 30 years of intensive toxicological investigation the important role of aflatoxin in cancer was revealed.

Owing to the heterogeneous distribution of aflatoxins in stored dietary staples and the monotonous diets of populations at risk of frequent and high levels of exposure, classical epidemiological methods were not ideal to distinguish variations in aflatoxin exposure levels within populations at risk. The use of validated exposure biomarkers allowed improved accuracy in such assessments at the individual level. The combination of animal carcinogenicity data and epidemiological data, based on exposure assessment using the validated biomarkers above, provided overwhelming evidence for aflatoxin B1 being a human liver carcinogen.[5] Readers are also directed to an excellent review of the animal and epidemiological data that established aflatoxins as important human carcinogens, which discusses genetic susceptibility factors in addition to synergistic interactions with hepatitis B virus, and further highlights intervention approaches to restrict exposure.[15]

6.2.6 Patterns of Aflatoxin Exposure in Children and the Perinatal Period

In countries with a high risk of exposure to aflatoxin the levels of aflatoxin–albumin in children are similar to those in adults from the same region; typically >95% have detectable aflatoxin–albumin adducts and within those individuals a wide range of adduct levels is frequently observed (50–200 fold), see Table 6.1.[37,38,40,42–45] Urinary aflatoxin M1 has also been observed frequently in young children from high risk countries, including Guinea, Sierra Leone and South Africa.[8,54,55] The presence of aflatoxins in cord blood collected at birth additionally indicates in utero exposure;[56–59] and further, the presence of aflatoxin–albumin[41,60] and aflatoxin–DNA adducts[61] in cord blood are highly suggestive of both transplacental aflatoxin transfer and that the fetus has the requisite metabolism to convert the toxin to its toxic and carcinogenic metabolites, the aflatoxin-epoxides. In one survey aflatoxin–albumin biomarkers were assessed in mothers during pregnancy, and subsequently in cord blood, and at week 16 and week 52 in the infant, see Table 6.1.[41,42] All maternal samples were positive, 49% of the cord blood samples, 11% of the week 16 and 92% of the week 52 samples were positive for aflatoxin–albumin. The ratio of maternal to cord blood adduct levels appeared to be about 10 : 1,[41,60] though not all maternal blood samples were collected at the time of birth. The frequency of aflatoxin–albumin detection at week 16 correlates with the initiation of weaning (about 20% at this time point).[41] The infant with the highest adduct level (30 pg/mg) at this time point was that of the earliest initiator of weaning, at 8 weeks, perhaps reflecting repeat exposures over the longest time frame. At 52 weeks the vast majority of infants had been started on weaning foods, and 92% of the infants had detectable aflatoxin–albumin adducts (mean 60.4 pg/mg; range non-detectable to 391 pg/mg).[42] Thus in countries at high risk for aflatoxin exposure, there appears to be only modest exposure gap for aflatoxin from the point of conception, through birth and throughout childhood and into adulthood.

6.2.7 Childhood Aflatoxin Exposure and Impaired Growth

The data from animals provide convincing evidence that aflatoxin exposure affects growth.[62] Given the early onset of chronic human exposure to aflatoxin at high levels in some regions, there is considerable concern over the potential effect of aflatoxin on growth. Studies in children have been suggestive of a link between aflatoxin and Kwashiorkor and/or malnutrition,[63,64] though study designs were not ideal, in part because validated exposure biomarkers, as described above, were not utilized and dose–responses were not observed. A survey of Gambian children aged 6–9 years revealed a modest inverse association between aflatoxin–albumin biomarker levels and both growth and a marker of immune susceptibility.[43] However, a stronger relationship between aflatoxin exposure and growth was revealed in younger children from Benin and Togo, who were aged less than 5 years old. Aflatoxin–albumin adducts were detected in 99% of blood samples (geometric mean 33 pg/mg, range, non-detectable to 1063 pg/mg), and a strong inverse association was found between aflatoxin–albumin adduct level and both stunting and being underweight ($p < 0.001$).[37] This study reported some of the highest levels of aflatoxin–albumin recorded; with 16% of samples exceeding the 100 pg/mg level and 1% with levels exceeding 500 pg/mg. By contrast, in the previous study of older Gambian children only 7% of samples were over 100 pg/mg and none was over 500 pg/mg. It is therefore not possible from these separate studies to assess whether the exposure range and/or the age range predominantly explain the differences in the observed strength of the relationship between aflatoxin exposure and growth faltering. A follow-up 8-month longitudinal study of 200 Beninese children aged 16–37 months confirmed this strong relationship between aflatoxin–albumin adduct levels and growth.[38] When the mean adduct levels within the children were divided into quartiles, the difference in aflatoxin–albumin adducts between the lowest and the highest group was 160 pg/mg and the difference in growth was 1.7 cm ($p < 0.001$, data were adjusted for height at recruitment and nutritional markers). Thus within this age group and within this exposure range a mean difference in aflatoxin exposure assessed using aflatoxin–albumin biomarkers of about 100 pg/mg approximates to about a 1 cm reduction in height within an 8-month period.

A more recent survey in Gambian women during pregnancy and their offspring in the first year of life revealed that *in utero* aflatoxin exposure (see Table 6.1 and text above) was associated with growth faltering during the first year of life.[41] There was a reduction in both growth velocity and weight gain in children associated with the level of the maternal aflatoxin–albumin adduct level assessed prior to birth. The poor growth velocity was further associated with the level of aflatoxin–albumin when the infants were 16 weeks old. An association between low birth weight and aflatoxin–albumin biomarkers in maternal blood collected post delivery has also been reported.[39] Figure 6.6 provides a summary of the main studies in which associations have been observed between the biomarker and early childhood growth.[37–39,41,43] The potential mechanisms of growth faltering are not clearly defined at this time, but they may in part reflect intestinal toxicity and/or immune modulation.[65,66]

It is likely that continued use of aflatoxin biomarkers will be important in terms of both liver cancer and growth faltering, in the applications listed below.

1. Understanding the synergistic interaction between aflatoxin and hepatitis B virus (HBV) in the aetiology of liver cancer. Preliminary data are suggestive that, in children, aflatoxin metabolism is modified by HBV

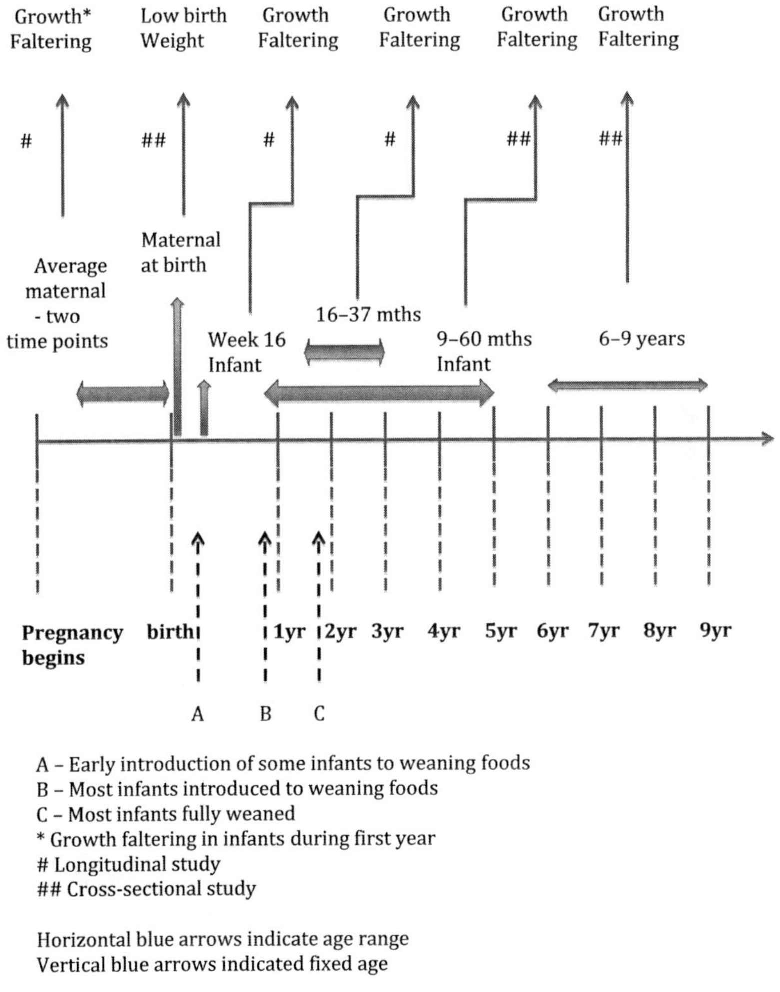

Figure 6.6 Summary for six studies of growth faltering data using aflatoxin biomarkers.[37–39,41,43]

infection, favouring the production of the highly reactive epoxide.[44] Understanding the differences in metabolite profiles between HBV carriers and non-carriers may be critical in this respect.
2. Understanding whether there is a critical period in the perinatal period for aflatoxin-induced growth faltering.
3. Understanding the mechanism of aflatoxin-induced growth faltering, including the potential and complex role of immune system modulation.
4. Monitoring intervention procedures to restrict aflatoxin exposure in at-risk populations.[15]

6.2.8 Aflatoxin Summary

In tropical parts of the world aflatoxin exposure through the natural contamination of dietary staples is common. The use of validated biomarkers has revealed that exposure is frequent and occurs at high levels in many of these populations. The use of validated exposure biomarkers was an essential component of the classification of aflatoxin B1 as a class 1A carcinogen.[5] In the last decade the use of these biomarkers is revealing strong associations between aflatoxin exposure and growth faltering.[37–39,41,42] Potential mechanisms of growth faltering have been reviewed,[66] and it is likely that the continued use of aflatoxin biomarkers will unravel critical mechanisms, and in turn assist sustainable, affordable and targeted intervention strategies to restrict exposure. In this respect it is worth emphasizing that approximately 4 billion individuals live in regions of the world that are at risk of aflatoxin exposure, through the "natural" contamination of their diet,[67] and therefore this undertaking is far from trivial.

6.3 Deoxynivalenol

6.3.1 Deoxynivalenol Structure

Deoxynivalenol (DON), also known as 12,13-epoxy-3,7,15-trihyroxy-trichothec-9-en-8-one; vomitoxin; dehydronivalenol; 4- deoxynivalenol or RD-toxin, is a common contaminant of cereals such as wheat, maize and barley.[1,2,5] It is part of a family of *Fusarium* mycotoxins known as the trichothecenes, and is a group B trichothecene (Figure 6.7).

6.3.2 Occurrence and Epidemiology of Deoxynivalenol

DON is a member of the trichothecene family of mycotoxins, predominantly associated with crop contamination by *Fusarium graminearum* (*Gibberella zeae*) and *F. culmorum* fungi, both of which are important plant pathogens that cause *Fusarium* head blight in wheat and *Gibberella* ear rot in maize.[68] The geographical distribution of the two species is related to temperature, and they are common in most temperate regions.[1,2,68–73] In developed countries, levels of DON are highly regulated and therefore "presumed" not to pose a significant

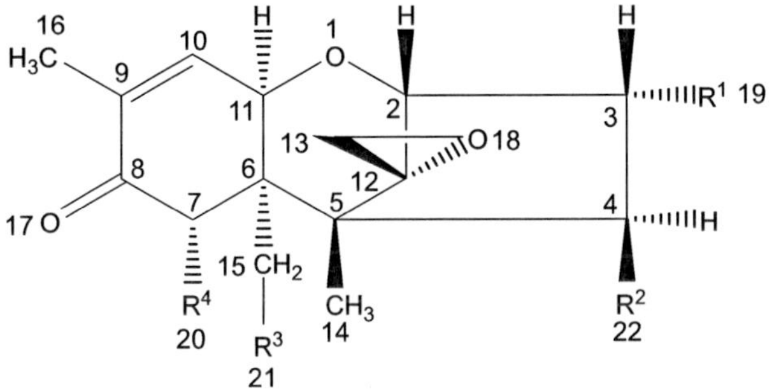

	R^1	R^2	R^3	R^4
Nivalenol	OH	OH	OH	OH
Deoxynivalenol	OH	H	OH	OH
3-Acetyl-Deoxynivalenol	$OCOCH_3$	H	OH	OH
15-Acetyl-Deoxynivalenol	OH	H	$OCOCH_3$	OH
Fusarenon X	OH	$OCOCH_3$	OH	OH

Figure 6.7 Generic structure of type B tricothecenes.[5]

risk to human health. In developing countries regulations are absent or poorly enforced, providing a greater possibility of more frequent exposure at high levels. To date it is clear that DON exposure is frequent in many countries, though the potential health effects of such exposures remain poorly investigated in both developed and developing regions.

During the twentieth century, mycotoxins were implicated in the aetiology of numerous food poisoning outbreaks, and DON was found frequently and often at high levels. Such observations were common in India and China,[69–71,74–77] and have been reviewed in detail.[77] The major and most frequent symptoms detected were similar to those in DON-exposed animals and included abdominal cramps, vomiting, diarrhoea, headaches and dizziness. In China, DON levels in some cases exceeded the Chinese regulatory limit of 1 ppm (or 1 μg/kg body weight/day) by >10 fold. Despite observations linking DON in foods to incidents of human gastroenteritis, confirmation of the aetiology in many outbreaks remains unknown. A number of poisoning incidences in US children are also suggestive of DON in their aetiology,[78] though this remains unconfirmed. In most cases inadequate sampling or analytical methods were used, inappropriate or no control subjects were included, and exposure biomarkers were not available. Novel biomarkers (see Section 6.3.3) are now revealing that DON exposure is frequent even in developed countries.[79,80]

6.3.3 Deoxynivalenol Toxicology

At a cellular and molecular level, different mechanisms have been proposed to explain or account for the toxicity of DON. The primary mechanism is its capacity to bind to eukaryotic ribosomes and inhibit protein synthesis. Indeed, studies conducted mainly in mice have shown that marked inhibition of protein synthesis occurs in most tissues,[68,77] although DON toxicity may also involve impaired membrane function, which alters intercellular communication and deregulates calcium homeostasis. Other studies suggest that early alterations in cell signalling mediated through mitogen activated protein kinase (MAPK)-dependent pathways are responsible for toxicity associated with DON exposure (reviewed by reference 77).

DON toxicity has been evaluated in a variety of different species with evidence of varying sensitivity. As previously documented,[81] swine are the most sensitive whilst ruminants are the least sensitive to DON exposure. With high-dose, short-term exposure to DON, acute or sub-acute toxicity is detected, characterized by histopathological effects such as haemorrhage of the intestinal tract, necrosis in bone marrow and lymphoid tissues, and lesions in the kidney and heart. In addition, a major feature is gastrointestinal (GI) tract toxicity and emetic effects, which may include abdominal distress, increased salivation, malaise, diarrhoea, emesis, and anorexia.[82] In contrast, prolonged intermediate to high dose exposure to DON causes decreased weight gain, anorexia and feed refusal.[83,84] Exposure to DON prior to conception and during pregnancy also has adverse effects on exposed animals. Symptoms such as reduced food intake in pregnant animals, weight loss, a decreased number of live fetuses and malformation of fetuses were reported in rat and mouse studies.[77]

At low levels of exposure DON can be immunostimulatory, influencing cytokine expression, positively or negatively, whilst at higher levels apoptosis in lymphoid tissues may lead to immunosuppression, or modify host resistance, humoral and cell-mediated responses, and serum IgA levels.[85–89]

6.3.4 Deoxynivalenol Metabolism and Biomarkers

In DON-contaminated crops the monoacetylated forms, 3-acetyl-deoxynivalenol (3AcDON) and 15-acetyl-deoxynivalenol (15AcDON), may additionally co-occur albeit at lower concentrations than DON; both are rapidly de-acetylated once ingested. Intestinal or ruminal microorganisms are responsible for de-acetylation, in addition to a de-epoxidation of DON to form the less toxic compound DOM-1.[90–94] The susceptibility to DON toxicity is in part related to the ability to de-epoxidate DON prior to uptake; DON-resistant species have gut or ruminal microbiota that are highly efficient in this step, while more sensitive species are less able to produce DOM-1.[77] A limited study by Eriksen and Pettersson revealed no de-epoxide metabolites in 10 faecal incubation mixtures from humans,[95] suggestive of a greater susceptibility.

A number of candidate exposure bio-measures have been suggested for DON, including the suppression of insulin-like growth factor acid labile subunit (IGFALS) and insulin growth factor 1 (IGF1) in mice,[87] although these and other inflammatory markers await further validation for sensitivity and specificity. The identification of biomarkers of DON exposure such as the parent compound itself and/or its metabolites may be more valuable for general biomonitoring purposes. A study conducted on the urine of exposed Sprague–Dawley male rats implied that the parent compound, free DON (fD), and the major metabolite DON–glucuronide (DG) combined (fD + DG) was a potentially useful biomarker for human exposure.[96] Further investigation by these authors allowed for the development of a robust urinary assay to detect individual exposure to DON using urinary fD + DG,[97,98] for which a strong correlation (adjusted $R^2 = 0.83$, $p < 0.001$) between DON intake (measured by food sampling over a four day period) and the urinary biomarker was subsequently demonstrated.[99] Urinary DON was also found to be stable in a cryostorage survey, albeit limited to date to only 3 years.[100] These are critical components of the validation process and thus there is now considerable confidence in the use of the combined measure of urinary fD + DG as an exposure biomarker for DON in epidemiological studies. To date surveys are somewhat limited in number and mostly restricted to European countries, though a survey in Egypt has recently been completed. Table 6.2 summarizes these observational data.[80,96–102]

Table 6.2 Urinary DON and DOM-1 biomarker measurements; fD, free DON; DG, DON–glucuronide (DG); DOM-1, de-epoxydeoxynivalenol.

	fD + DG pos/ total (%)	*Mean[a] fD + DG ng/ml (range)*	*DOM-1 Pos/total*	*Mean DOM-1 ng/ml (range)*	*Ref.*
Normal diet					
UK	296/300 (99)	9 (nd–59)	not tested	/	98
UK	25/25 (100)	7 (1–59)	0/22 (0)	0	97
UK[d]	35/35 (100)	9 (1–49)	1/34 (3)	1	99
	34/35 (97)	8 (nd–59)	0/35 (0)	0	
	31/35 (86)	9 (nd–78)	1/35 (0)	<1	
	34/35 (97)	12 (nd–49)	0/35 (0)	0	
	32/35 (91)	10 (nd–58)	0/35 (0)	0	
	33/35 (94)	9 (nd–62)	0/35 (0)	0	
UK	85/86 (99)	10 (nd–117)	0/86 (0)	0	102
France	75/76 (99)	7[b] (nd–29)	26/76 (34)	<1[b] (nd–3)	100
Egypt	63/93 (68)	3[c] (nd–60)	2/93 (2)	both <1	101
Sweden	28/29 (97)	11 (nd–66)	2/29 (7)	<1 and 1.0	80
China	15/15 (100)	30 (4–94)	not tested	/	96
Restricted diet[e]					
UK	9/25 (36)	<1 (nd–8)	not tested	/	97
UK[d]	6/10 (60)	<1 (nd–3)	not tested	/	99
	5/10 (50)	<1 (nd–3)	not tested	/	
	3/10 (30)	<1 (nd–2)	not tested	/	
	2/9 (22)	<1 (nd–1)	not tested	/	

[a] Geometric means presented.
[b] Median of positives.
[c] Geometric means of positives.
[d] Longitudinal data over 6 days (normal diet) and 4 days (restricted diet).
[e] Dietary intervention to restrict major sources of wheat *e.g.* bread, pasta, cakes, biscuits, breakfast cereals.

DG is the major metabolite in human urine, and the limited data to date indicate that within UK the average ratio of fD to DG was 1 : 9, though it varied by individual.[103] This ratio appears to track within individuals when comparing samples on different days and at different exposure levels, thus the ratio may be a useful phenotypic marker of susceptibility to DON toxicity. To date DOM-1 is either absent or infrequently observed in the urine, and where it is present it occurs at only modest levels (see Table 6.2). To date urinary DOM-1 has not been validated as an exposure biomarker for DOM-1 production, thus some caution in interpretation is required, but the lack or low levels of urinary DOM-1 in any study is strongly supportive that humans may be a particularly sensitive species to DON toxicity.[80,95,97,99,100,102]

On the basis of the concentration of the urinary biomarker and the average urinary transfer of DON, determined by measures of DON intake and the urinary biomarker at ~75%,[100] the mean intake of DON for adults was estimated to be 200–250 ng/kg body weight (bw)/day, with between 1 and 5% of participants predicted to exceed the European Union recommended PMTDI of 1000 ng DON/kg bw/day.[104] Higher levels of exposure are predicted outside of the EU.

New "omics" approaches have also been applied to DON biomarker research. One study reported gene expression profiles that are indicative of exposure to immunotoxicants, such as DON, by investigating whole-genome gene expression in peripheral blood mononuclear cells (PBMCs). A set of 48 genes was identified that distinguished the immunotoxic from the non-immunotoxic compounds, providing a promising transcriptomic profile indicative of immunotoxic exposure.[105] Proteomics has also been applied to DON biomarker research with promising results. First, studies reporting the effects of low dose DON exposure in human lymphocyte cell lines, utilizing two-dimensional gel electrophoresis coupled to mass spectrometry (MS), identified highly reproducible quantitative and phosphorylation changes to the proteome. The candidate protein biomarkers provide insights as to how low dose exposure to DON may affect human immune function and may provide mechanism-based biomarkers for DON exposure.[106,107] Second, a study on the mouse thymoma cell line EL4 investigated changes to protein expression, utilizing a one-dimensional sodium dodecylsulfate polyacrylamide gel electrophoresis (SDS-PAGE) approach, demonstrated that the up-regulation of transcription factor My-binding protein 1a (a known repressor of transcription factors such as NF-kB) might play a role in the mechanism of DON toxicity.[108] A urinary metabolomics examination of human urine samples also revealed a 1.5-fold increase in hippurate in individuals exposed to high levels of DON compared to low levels,[109] though this study awaits additional larger field trials for confirmation.

6.3.5 Deoxynivalenol Summary

DON exposure assessment is of growing interest owing to its possible effects on human health. Within the European Union it was the most frequent contaminant of the cereals maize and wheat. The biomarker driven data, predominantly from the UK at this time, indicate that exposure is frequent, occurring in >90% of tested individuals, with some likely to exceed the recommended tolerable daily intake. Exposures outside of Europe, with perhaps the exception of Africa, are likely to be higher than in the UK. At a cellular level, DON has been associated with disrupting cell signalling, differentiation, growth and macromolecular synthesis. Given that DON influences immune function in animals, it is also likely to exert immunomodulatory effects in humans. The main challenges in biomarker discovery are the identification and confirmation of such candidate biomarkers *in vitro* and the subsequent validation and application *in vivo*. The pursuit of an individual biomarker to justify a specific exposure or effect may not prove fruitful. Instead, the pursuit of a combination of biomarkers may be more appropriate and contribute to an understanding of the toxicity mechanism associated with the exposure and/or effect of a specific stressor. Novel technologies, such as the "omics", provide an invaluable new approach to the development of mechanism-based biomarkers for DON exposure, because they present the opportunity to examine a large number of molecular signatures in a tissue or cell of interest and thereby contribute to an understanding of the effects of DON through molecular epidemiology studies. The research arena eagerly awaits the use of molecular epidemiological derived biomarkers to assess the potential health effects. The pilot exposure data on DON from a major mother birth cohort, "Born in Bradford", may be an important resource to conduct such studies.[102]

6.4 Fumonisins

6.4.1 Fumonisins Structure

Fumonisins are secondary metabolites produced predominantly by *Fusarium verticillioides* (Sacc.) Nirenberg (formerly known as *F. moniliforme* Sheldon) and *F. proliferatum* (Matsushima) Nirenberg.[110] At least 28 fumonisin analogues have been described and categorized into A, B, C, and P series; fumonisin B_1, B_2 and B_3 (FB_1, FB_2 and FB_3) are the most abundant naturally occurring fumonisins, of which FB_1 (see Figure 6.8) is the most significant analogue, usually dominating at >70% of the total fumonisins detected in natural maize samples.

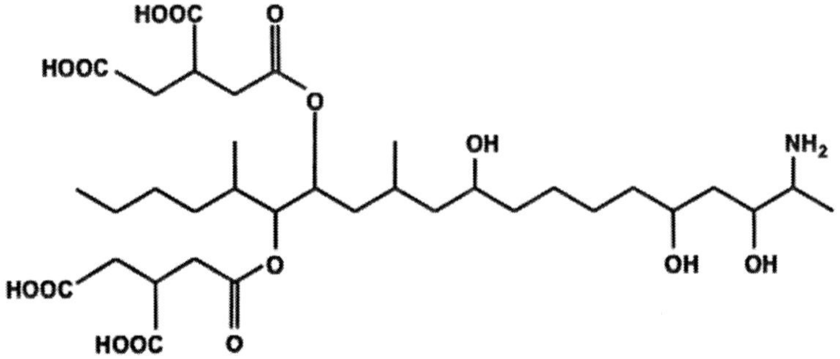

Figure 6.8 Structure of Fumonisin B1.[5]

6.4.2 Fumonisin Toxicology

Fumonisins cause various distinct syndromes in different animals, such as leukoencephalomalacia in horses, pulmonary oedema in pigs and neural tube defects in mice, reviewed by references.[110,111] Although fumonisins are not mutagenic or genotoxic in primary rat hepatocytes, FB_1 exhibits clastogenesis and epigenetic properties in cell culture. FB_1 is hepatocarcinogenic in male BD IX rats and in B6C3F1 female mice, and nephrocarcinogenic in male Fischer 344 rats.[112–114] High levels of fumonisins have been found in naturally contaminated maize from areas with a high incidence of oesophageal cancer, *e.g.* Centane magisterial district, Eastern Cape Province, South Africa; Cixian County, Hebei Province, China; Mazandaran Province, Iran; State of Santa Catarina, Brazil; South Carolina, USA; and Northern Italy.[5,113,115] Fumonisins have also been associated with primary liver cancer in Haimen, Jiangsu Province, China.[116] Based on current data, the International Agency for Research on Cancer has classified FB_1 to be possibly carcinogenic to humans, group 2B.[5] In addition, the consumption of maize contaminated with fumonisins has been reported as a risk factor for human neural tube defects (NTDs), such as in the Eastern Cape Province of South Africa; the Northern provinces of China; and along the Texas–Mexico border in North America.[117,118] FB_1 inhibits 5-tetramethylfolate uptake via complex sphingolipid depletion, thus folate deficiency is postulated as a mechanism of fumonisin-induced NTDs.[119]

6.4.3 Predictors and Identification of High Risk Populations

In developed countries maize forms a minor part of the diet. Maize intake is estimated at less than 10 g/person/day in the European Union and the maize that is consumed tends to be of good quality.[120] Generally in developing countries, and more specifically in certain rural areas, maize forms a progressively larger part of the diet, almost to the exclusion of all other food commodities.[113] In rural communities in South America and South Africa average maize consumption is estimated at 250 and 460 g/person/day, respectively.[121–123] Subsistence farming communities that consume maize as a staple diet can be exposed to total fumonisin levels as high as 12.1 μg/kg bw/day (95% CI: 0.3–4927).[123] This is of considerable concern because the FAO/WHO Expert Committee on Food Additives (JECFA) has determined a group provisional maximum tolerable daily intake (PMTDI) for fumonisin B_1, B_2 and B_3, alone or in combination, of 2 μg/kg bw/day.[120]

6.4.4 Fumonisin Toxicokinetics, Metabolism and Biomarkers

Toxicokinetic studies have shown that fumonisins are poorly absorbed and have a very low bioavailability.[124] FB_1 has an elimination half-life of less than an hour when administered via different routes such as by gavage, intravenously or intraperitoneally in various animal studies.[125–131] The toxicokinetics of FB_2 in rats and vervet monkeys are similar to those of FB_1.[132,133] The lower levels of FB_2 detected in the urine of both rats and monkeys dosed by intravenous and by interperitoneal injection can be ascribed to the lower polarity of FB_2 compared to FB_1. By extrapolating data reported in various animal studies, it was estimated that the half-life in a 70-kg human would be 128 minutes, based on the allometric relationship between animal weight and the toxicokinetic half-life of the FB_1 administered.[122] In vervet monkeys and swine between 0.3 and 2% of the FB1 dose is excreted in urine.[129,131,134]

Fumonisins exert their main biochemical effect by inhibiting ceramide synthase, a key enzyme in the *de novo* sphingolipid biosynthetic pathway, preventing the conversion of sphinganine to dihydroceramide and the re-acylation of sphingosine to ceramide.[135,136] The disruption of the sphingolipid biosynthetic pathway elevates the levels of the sphingoid bases and their 1-phosphates and decreases ceramide and more complex sphingolipids, such as sphingomyelin and gangliosides, and their intermediates. Sphingolipids are predominantly found in cellular membranes and are critical for the maintenance of the membrane structure, while complex sphingolipids function as precursors for second messengers and are important in sustaining cellular growth and differentiation.[135] The disruption of sphingolipid metabolism by fumonisins causes sphinganine, and sphingosine to a lesser extent, to increase, and the resultant increase in the sphinganine/sphingosine ratio has been proposed as a biomarker of fumonisin exposure.[136] Various animal studies have successfully demonstrated the potential of the sphinganine/sphingosine ratio as a biomarker of fumonisin exposure in serum and urine.[136–140] Sphinganine and sphingosine, as well as their ratio, have also been investigated in several human studies in blood and urine, but despite a range of Sa/So being observed, no consistent correlation between the bio-measurement and estimated fumonisins ingestion are reported (Table 6.3a/b).[141–147] The use of the ratio may be problematic because of the potential for adaption to previous fumonisin exposure, the possible higher and more extended responses to FB2 compared to FB1, and the possible confounding by co-exposure to aflatoxin, as reviewed previously.[141] In some studies less than ideal study design has also been problematic. Overall, neither sphinganine, spingosine, nor their ratio, have proved to be successful biomarkers of human exposure to fumonisins.

In animal studies most fumonisins were excreted almost unchanged in the faeces and only a small percentage was excreted in urine.[125–131] Given the lack of significant FB metabolism, urinary FB_1 was suggested as a putative biomarker of exposure in human urine.[148] Urinary FB1 (range 19–248 pg/mL, n = 75) was associated with consumption of maize-based tortillas by Mexican women in whom samples were selected from a larger cohort to represent low, medium and high tortilla consumption groups, 25 women per group.[149] Fumonisins have been implicated in the aetiology of oesophageal cancer, and median urinary FB1 levels were approximately 10 times higher in healthy individuals in Huaian (a high risk oesophageal cancer area) compared to Fusui (a high risk hepatocellular cancer area), although no correlation between estimated intake of fumonisins and the urinary measure was found.[116] A laboratory driven sorting and hand washing intervention to restrict the levels of

Table 6.3a Exposure assessment data and sphinganine (Sa)/sphingosine (So) ratio in plasma as a putative biomarker of fumonisin exposure.

Plasma/Serum	*n*	*Fumonisin exposure (μg/kg bw/day)*	*Sphinganine (nM)*	*Sphingosine (nM)*	*Sphinganine/ Sphingosine*	*References*
France, healthy volunteers	17				0.4 ± 0.2	139
South Africa, healthy volunteers	13				0.2 ± 0.1	139
South Africa, cancer patients	4				0.2 ± 0.1	139
Centane, South Africa	154	3.8	20 ± 28	75 ± 83	0.3 ± 0.4	145
KwaZulu-Natal, South Africa	27	0	67 ± 88	128 ± 115	0.4 ± 0.2	145
Bomet, Kenya	29	0.1	59 ± 31	215 ± 105	0.3 ± 0.1	145
Centane, South Africa	152	4.4	18 ± 25	75 ± 84	0.3 ± 0.3	146
Bizana, South Africa	150	5.8	10 ± 10	56 ± 35	0.2 ± 0.2	146
Burkina Faso	62	0.8 (<0.1–2.40)	11 ± 4.8	19 ± 18	0.6 (range 0.1–3.0)	141
Centane, South Africa	36	4.1 ± 7.6	7.0 ± 6.3	22 ± 14	0.3 ± 0.1	147
Bizana, South Africa	30	3.9 ± 7.3	5.6 ± 2.5	21 ± 10	0.3 ± 0.1	147
Huaian, China	43	7.7	8.2 ± 3.7	20 ± 10	0.5 ± 0.2	116
Fusui, China	34	2.1	15 ± 5.4	18 ± 5.5	0.8 ± 0.1	116

Table 6.3b Exposure assessment data and Sa/So in urine as a putative biomarker of fumonisin exposure.

Urine	*n*	*Fumonisin exposure (μg/kg bw/day)*	*Sphinganine (nM)*	*Sphingosine (nM)*	*Sphinganine/ Sphingosine*	*References*
France, Females	17		0.4–7.0	0.9–17	<0.1–0.6	139
Centane, South Africa	153	3.8	5.1 ± 7.8	21 ± 29	0.4 ± 0.7	145
KwaZulu-Natal, South Africa	27	0	1.5 ± 2.0	6.8 ± 7.4	0.3 ± 0.2	145
Henan, China (Pre-harvest)	28	0			0.2 (<0.1–0.2)	142
Henan, China (Post-harvest)	28	184 (0.5–740)			0.2 (<0.1–0.9)	142
Southern Brazil	116	0.6			1.6 ± 0.5	143
Northern Argentina	74	0.6			0.7 ± 0.1	143
Central Argentina	20				0.4 ± <0.1	143
Southern Italy	66				0.4 ± <0.1	143
Burkina Faso	87	0.8 (<0.1–2.40)	76 ± 135	263 ± 327	0.4 (<0.1–7.2)	141
Centane, South Africa	152	4.4	4.6 ± 7.0	22 ± 30	0.3 ± 0.3	146
Bizana, South Africa	150	5.8	4.7 ± 6.1	15 ± 20	0.3 ± 0.1	146
Centane, South Africa	36	4.1 ± 7.6	3.4 ± 4.6	15 ± 20	0.2 ± 0.1	147
Bizana, South Africa	30	3.9 ± 7.3	7.1 ± 14	31 ± 63	0.2 ± 0.1	147
Huaian, China	43	7.7	2.9 ± 2.2	8.5 ± 4.9	0.4 ± 0.2	116
Fusui, China	34	2.1	14 ± 28	4.7 ± 6.7	0.7 ± 0.5	116
Portugal (Urban)	6	0.2	1.3 ± 3.4	3.0 ± 6.7	0.4 ± 0.2	144
Portugal (Rural)	9		0.8 ± 1.5	2.3 ± 4.3	0.4 ± 0.2	144

Table 6.4 Exposure assessment data and free fumonisin measures as putative fumonisin biomarkers in humans.

Urine	*n*	*Fumonisin exposure (μg/kg bw/day)*	*Maize intake (g/day)*	*Urinary FB_1 (pg/mg creatinine)*		*References*
Centane, South Africa	178	8.7 ± 0.2	456 ± 12			137
Bizana, South Africa	229	3.4 ± 0.2	380 ± 11			137
Mexico (Highest consumption group)				134 (95% CI: 79, 228)		149
Centane, South Africa (Baseline)		6.7 (95% CI: 1.2, 2.3)	340 (95% CI: 280, 400)	470 (95% CI: 295, 750)		152
Centane, South Africa (Intervention)		2.6 (95% CI: 1.9, 3.4)	380 (95% CI: 310, 470)	279 (95% CI: 202, 386)		152
Huaian, China	43	7.7	Median 200	Median 3900		116
Fusui, China	34	2.1	Median 250	Median 400		116
Faeces	*n*	*Fumonisins in maize (mg/kg)*		*Feacal FB_1 (mg/kg)*		*References*
KwaZulu-Natal (Rural), South Africa		2.2 (range 0.1–22.2)		9.0 (range 0.5–39)		154
KwaZulu-Natal (Urban), South Africa		0.3 (range 0.2–0.5)		6.8 (range 0.6–16)		154
Buccal Cells	*n*	*Fumonisin exposure (μg/kg bw/day)*	*Sphinganine (nM)*	*Sphingosine (nM)*	*Sphinganine/ Sphingosine*	*References*
Burkina Faso	64	0.8 (range <0.1–2.40)	12 ± 16	30 ± 37	0.5 (range 0.1–1.9)	141
Hair (Composite sample)	*n*			*FB_1 in hair (μg/kg)*		
Transkei, South Africa				22–33 μg/kg		153

fumonisins in maize kernels for cooking[150] was successfully applied to a field setting in the Centane magisterial district of South Africa, where a significant reduction in both the mean maize contamination with fumonisins and the levels of the bio-measure, urinary FB_1, were recorded.[151,152] The transfer of ingested FB_1 to urinary output was estimated to be less than 0.1%. Human hair and faeces have also been suggested as a possible matrix for use to develop an exposure biomarker for fumonisins.[153,154] FB1 bio-measures are summarized in Table 6.4.

6.4.5 Fumonisin Summary

Fumonsins are frequent contaminants of maize, and individuals at particular risk of exposure tend to be poorer farming communities in tropical regions, whose diet is predominantly maize. Such populations may also be at risk of aflatoxin exposure, though again this varies by geographical location. Fumonisins are classified as possibly carcinogenic, and implicated in the aetiology of neural tube defects. Investigations of the disruption of sphingolipid metabolism have not yielded a useful or viable exposure biomarker; urinary FB1 by contrast is a strong candidate, and significant advances have been made in the last three years in the development of a validated exposure biomarker. Significant advances in food safety and health in subsistence maize farming communities exposed to high levels of fumonisin could be possible by further development of the simple intervention approach described above.

6.5 Mycotoxin Biomarkers Summary

Mycotoxins are frequent contaminants of cereal crops throughout the world. The heterogeneous contamination and the homogeneous diets in regions at greatest risk of exposure lead to chronic exposure at high levels in many developing countries. Moderate chronic exposure to *Fusarium* mycotoxins in more temperate regions is also apparent. Mycotoxins can cause gross acute toxicity affecting hundreds to thousands of individuals on occasions; and they can be fatal at high levels. Some mycotoxins are classified as carcinogenic or possibly carcinogenic, thus chronic exposure to aflatoxin for example is a major contributor to the burden of liver cancer. The use of aflatoxin biomarkers was critical to confirm this assessment. An even greater burden of morbidity, and perhaps mortality, is now predicted from studies using aflatoxin biomarkers to assess immune effects and growth; though these more subtle affects are perhaps more complex to quantify. Other mycotoxins including those discussed here are likely to contribute to this disease burden, and biomarkers of exposure and effect will contribute to our understanding of the molecular epidemiology of mycotoxins in chronic disease. They will also support our understanding of approaches to intervene to restrict exposure. Table 6.5 summarizes the bio-measures and biomarkers now in place, and Table 6.6 provides an approximate time line for mycotoxin isolation and biomarker development.

Table 6.5 Summary of exposure bio-measures and biomarkers. AF-alb, aflatoxin–albumin; AFM1, aflatoxin M1; AF-N7-Gua, aflatoxin N7-guanine; FB1, fumonisin B1; fD + DG, free DON plus DON-glucuronide; DOM-1, de-epoxydeoxynivalenol; Sa/So – spinganine/spingosine ratio.

	Matrix	*Dose transferred*	*Relevant time frame*	*Validated*
Aflatoxin[b]				
AFM1	Urine	1–3%	24–48 h	Yes
AF-N7-Gua	Urine	1%	24–48 h	Yes
AFM1	Breast milk	0.1–2%	n/e	No
AF-alb	Serum/plasma	1–3%	2–3 months	Yes
Deoxynivalenol[c]				
fD + DG	Urine	75%	24–48 h	Yes
DOM-1	Urine	<5%	24–48 h[a]	No
Fumonisin[d]				
Sa/So	Urine	Not relevant	n/e	No
Sa/So	Serum	Not relevant	n/e	No
FB1	Urine	<1%	24–48 h[a]	No (strong candidate)

[a] predicted but not demonstrated.
[b] Based on references 5,7,15,47,66,115.
[c] Based on references 85,87.
[d] Based on references 115,137,149,152.

Key Points

- Mycotoxins occur naturally in the diet; avoidance of all types is unlikely to occur without a highly restrictive diet.
- Geographical, climatic and financial differences in world regions are important determinants of exposure level and exposure frequency.
- Animal data concerning the toxicity of mycotoxins are overwhelming, while mycotoxin links to human disease are scarce, with the exception of the aflatoxins.
- The development and use of validated exposure biomarkers was critical in understanding the role of aflatoxins in the aetiology of primary liver cancer.
- Aflatoxin biomarkers will be critical in understanding the role and extent of aflatoxins in growth faltering and immune toxicity. This is a problem likely to exceed that of liver cancer cases by one or more orders of magnitude.
- Use of aflatoxin biomarkers will continue to support critical efforts to restrict dietary exposures to aflatoxin in high risk settings.
- Urinary deoxynivalenol was strongly correlated with DON intake, and thus this exposure biomarker awaits application in epidemiological studies. Growth faltering and immune effects will probably dominate these studies.

Table 6.6 Approximate time frame by decade of mycotoxin discovery, biomarker development and health outcomes.

	1960s and 70s	*1980s*	*1990s*	*2000s*	*2010s*
Aflatoxin reviewed by 5, 7,15,47,66,115	Isolated (1961) Urinary AFM1 measured 1973	Urinary DNA adduct (1982) and serum AF–albumin adduct (1987) identified	Urinary AFM1, AF-N7-Gua, and serum AF–alb validated biomarkers AFB1 classified as human carcinogen, and, synergistic interaction with hepatitis B AF biomarkers used in intervention studies	AF biomarkers used in intervention studies Use of AF biomarkers reveals novel associations with health (growth and immunity)	Continued use of AF to understand novel health outcomes and implement interventions Anticipate evaluation of multi-mycotoxin synergistic interactions, and HBV interactions
Deoxynivalenol reviewed by 5,68,77,79, 80,85,87,99	Isolated (1972)	Ecological studies suggests links to oesophageal cancer	Acute poisoning incidents implicate DON	Early development of urinary DON biomarkers	Validation of urinary DON as an exposure biomarker[a] IGF suggested as biomarker of effect
Fumonisin reviewed by 5,110,115, 137,149,152		Ecological studies suggestive of links between *Fusarium* and oesophageal cancer FBs isolated (1988)	FB implicated in the aetiology of oesophageal cancer	Shinganine & sphingosine investigated as putative biomarkers Urinary FB1 investigated as putative biomarker	Urinary FB strong candidate as an exposure biomarker[a]

[a]Anticipate biomarker use in epidemiological studies.

- Urinary fumonisins are the strongest candidate for an exposure biomarker for fumonisin, though given the low transfer rate dose–response effects and application are likely only in very high risk settings, which in many ways is ideal.
- Given the probable more subtle effects of deoxynivalenol, fumonisins and other important fungal toxins, epidemiological studies of exposure biomarkers are likely to be complemented with effect biomarkers, which will allow a mechanistic understanding of the process. Such a paradigm is already in place for aflatoxins and liver cancer, reviewed by Kensler and colleagues.[15]

References

1. Council for Agricultural Science and Technology (CAST), Potential economic costs of mycotoxins in the United States, in *Mycotoxins: Risks in Plant, Animal and Human Systems,* Task force report No. 139, Ames, IA, USA, 2003, pp. 136–142.
2. J. D. Miller, Fungi and mycotoxins in grain: implications for stored product research, *J. Stored Prod. Res.*, 1995, **31**, 1–16.
3. P. M. Scott, Biomarkers of human exposure to ochratoxin A, *Food Add. Contam.*, 2005, **22** Suppl 1, 99–107.
4. J. Gilbert, P. Brereton and S. MacDonald, Assessment of dietary exposure to ochratoxin A in the UK using a duplicate diet approach and analysis of urine and plasma samples, *Food Add. Contam.*, 2001, **18**(12), 1088–1093.
5. IARC, Monograph 56, Some naturally occurring substances: food items and constituents, heterocyclic aromatic amines and mycotoxins, International Agency for Research on Cancer, Lyon, France, 1993.
6. A. A. Stark, Threat assessment of mycotoxins as weapons: molecular mechanisms of acute toxicity, *J. Food Prot.*, 2005**, 68**(6), 1285–1293.
7. C. P Wild and P. C. Turner, The toxicology of aflatoxins as a basis for public health decisions, *Mutagenesis*, **17**(6), 471–481.
8. N. Polychronaki, C. P. Wild, H. Mykkänen, H. Amra, M. Abdel-Wahhab, A. Sylla, M. Diallo, H. El-Nezami and P. C. Turner, Urinary biomarkers of aflatoxin exposure in young children from Egypt and Guinea, *Food Chem. Toxicol.*, 2008, **46**, 519–526.
9. N. Polychronaki, R. M. West, P. C. Turner, H. Amra. M. Abdel-Wahhab, H. Mykkänen and H. El-Nezami, A longitudinal assessment of aflatoxin M_1 excretion in breast milk of selected Egyptian mothers, *Food Chem. Toxicol.*, 2007, **45**, 1210–1215.
10. N. Polychronaki, P. C. Turner, H. Mykkänen, Y. Gong, H. Amra. M. Abdel-Wahhab and H. El-Nezami, Determinants of aflatoxin M_1 in breast milk of Egyptian mothers, *Food Add. Contam.*, 2006, **23**(7), 700–708.

11. P. C. Turner, C. Loffredo, S. El-Kafrawy, S. Ezzat, S. Eissa, S. Abdel-Latif, M. El-Daly, O. Nada and M. Abdel-Hamid, A survey of aflatoxin-albumin adducts in sera from Egypt, *Food Add. Contam.*, 2008, **5**, 583–587.
12. V. M. Scussel, P. Haas,Y. Gong, P. C. Turner and C. P. Wild, Study of aflatoxin exposure in a Brazilian population using an aflatoxin-albumin biomarker, in *Mycotoxins and Phycotoxins: Advances in Determination, Toxicology and Exposure Management*, ed. H. Njapau, S. Trujillo, H. P van Egmond and D. L. Park, Wageningen Academic Press, 2006, pp. 197–202.
13. C. P. Wild, Y. Z. Jiang, S. J. Allen, L. A. Jansen, A. J. Hall and R. Montesano, Aflatoxin-albumin adducts in human sera from different regions of the world, *Carcinogenesis*, 1990, **11**(12), 2271–2274.
14. P. C. Turner, C. P. Wild and P. Scott. Aflatoxin-albumin biomarker survey of 200 Canadian sera samples (unpublished observation).
15. T. W. Kensler, B. D. Roebuck, G. N. Wogan and J. D. Groopman. Aflatoxin: A 50 year odyssey of mechanistic and translational toxicology, *Toxicol. Sci.*, 2010. [Epub ahead of print] doi: 10.1093/toxsci/kfq283.
16. F. P. Guengerich, Y. F. Ueng, B. R. Kim, S. Langouet, B. Coles, R. S. Iyer, R. Thier, T. M. Harris, T. Shimada, H. Yamazaki, B. Ketterer and A. Guillouzo, Activation of toxic chemicals by cytochrome P450 enzymes: regio- and stereoselective oxidation of aflatoxin B1, *Adv. Exp. Med. Biol.*, 1996, **387**, 7–15.
17. E. P. Gallagher, K. L. Kunze, P. L. Stapleton and D. L. Eaton, The kinetics of aflatoxin B1 oxidation by human cDNA-expressed and human liver microsomal cytochromes P450 1A2 and 3A4, *Toxicol. Appl. Pharm.*, 1996, **141**(2), 595–606.
18. M. Kitada and T. Kamataki, Cytochrome P450 in human fetal liver: significance and fetal-specific expression, *Drug Metab. Rev.*, 1994, **26**, 305–323.
19. D. Lacroix, M. Sonnier, A. Moncion, G. Cheron and T. Cresteil, Expression of CYP3A in the human liver – evidence that the shift between CYP3A7 and CYP3A4 occurs immediately after birth, *Eur. J. Biochem.*, 1997, **247**(2), 625–34.
20. S. C. Sim, R. J. Edwards, A. R. Boobis and M. Ingelman-Sundberg, CYP3A7 protein expression is high in a fraction of adult human livers and partially associated with the CYP3A7*1C allele, *Pharmacogen. Genomics*, 2005, **15**(9), 625–31.
21. D. L. Eaton, H. S. Ramsdell and G. E. Neal, Biotransformation of the aflatoxins, in *The Toxicology of Aflatoxins: Human Health, Veterinary and Agricultural Significance*, ed. D. L. Eaton and J. D. Groopman, Academic Press, 1994, pp. 45–72.
22. V. M. Raney, T. M. Harris and M. P. Stone, DNA conformation mediates aflatoxin B1-DNA binding and the formation of guanine N7 adducts by aflatoxin B1 8,9-exo-epoxide, *Chem. Res. Toxicol.*, 1993, **6**(1), 64–68.
23. J. D. Groopman, A. Hall, H. Whittle, G. Hudson, G. N. Wogan, R. Montesano and C. P. Wild, Molecular dosimetry of aflatoxin-N7-guanine in human urine obtained in The Gambia, West Africa, *Cancer Epidemiol. Biomarkers Prev.*, 1992, **1**, 221–228.

24. J. D. Groopman, J. Q. Zhu, P. R. Donahue, A. Pikul, L. S. Zhang, J. S. Chen and G. N. Wogan, Molecular dosimetry of urinary aflatoxin-DNA adducts in people living in Guangxi Autonomous Region, People's Republic of China, *Cancer Res.*, 1992, **52**(1):45–52.
25. F. P. Guengerich, W. W. Johnson, T. Shimada, Y. F. Ueng, H. Yamazaki and S. Langouët, Activation and detoxication of aflatoxin B1, *Mutat. Res.*, 1998, **402**(1-2), 121–128.
26. S. Bodreddigari, L. K. Jones, P. A. Egner, J. D. Groopman, C. H. Sutter, B. D. Roebuck, F. P. Guengerich, T. W. Kensler and T. R. Sutter, Protection against aflatoxin B1-induced cytotoxicity by expression of the cloned aflatoxin B1-aldehyde reductases rat AKR7A1 and human AKR7A3, *Chem. Res. Toxicol.*, 2008, **21**(5), 1134–1142.
27. F. P. Guengerich, H. Cai, M. McMahon, J. D. Hayes, T. R. Sutter, J. D. Groopman, Z. Deng and T. M. Harris, Reduction of aflatoxin B1 dialdehyde by rat and human aldo-keto reductases, *Chem. Res. Toxicol.*, 2001, **14**(6), 727–737.
28. D. P. H. Hsieh and J. J. Wong, Biotransformation of the aflatoxins, in *The Toxicology of Aflatoxins: Human Health, Veterinary and Agricultural Significance*, ed. D. L. Eaton and J. D. Groopman, Academic Press, 1994, pp. 73–88.
29. G. Sabbioni, P. L. Skipper, G. Büchi and S. R. Tannenbaum, Isolation and characterization of the major serum albumin adduct formed by aflatoxin B1 in vivo in rats, *Carcinogenesis*, 1987, **8**(6), 819–824.
30. C. P. Wild, G. J. Hudson, G. Sabbioni, B. Chapot, A. J. Hall, G. N. Wogan, H. Whittle, R. Montesano and J. D. Groopman, Dietary intake of aflatoxins and the level of albumin-bound aflatoxin in peripheral blood in The Gambia, West Africa, *Cancer Epidemiol. Biomarkers Prev.*, 1992, **1**(3), 229–234.
31. L. S. Gan, P. L. Skipper, X. C. Peng, J. D. Groopman, J. S. Chen, G. N. Wogan and S. R. Tannenbaum, Serum albumin adducts in the molecular epidemiology of aflatoxin carcinogenesis: correlation with aflatoxin B1 intake and urinary excretion of aflatoxin M1, *Carcinogenesis*, 1988, **9**(7), 1323–1235.
32. G. Sabbioni, S. Ambs, G. N. Wogan and J. D. Groopman, The aflatoxin-lysine adduct quantified by high-performance liquid chromatography from human serum albumin samples, *Carcinogenesis*, 1990, **11**(11), 2063–2066.
33. P. F. Scholl and J. D. Groopman, Long-term stability of human aflatoxin B1 albumin adducts assessed by isotope dilution mass spectrometry and high-performance liquid chromatography-fluorescence, *Cancer Epidemiol. Biomarkers Prev.*, 2008, **17**(6), 1436–1439.
34. L. F. McCoy, P. F. Scholl, A. E. Sutcliffe, S. M. Kieszak, C. D. Powers, H. S. Rogers, Y. Y. Gong, J. D. Groopman, C. P. Wild and R. L. Schleicher, Human aflatoxin albumin adducts quantitatively compared by ELISA, HPLC with fluorescence detection, and HPLC with isotope dilution mass spectrometry, *Cancer Epidemiol. Biomarkers Prev.*, 2008, **17**(7), 1653–1657.
35. P. F. Scholl, P. C. Turner, A. E. Sutcliffe, A. Sylla, M. Diallo, M. D. Friesen, J. D. Groopman and C. P. Wild, Quantitative comparison of

aflatoxin B1 serum albumin adducts in humans by isotope dilution mass spectrometry and ELISA, *Cancer Epidemiol. Biomarkers Prev.*, 2006, **15**(4), 823–826.

36. N. M. Johnson, G. Qian, L. Xu, D. Tietze, A. Marroquin-Cardona, A. Robinson, M. Rodriguez, L. Kaufman, K. Cunningham, J. Wittmer, F. Guerra, K. C. Donnelly, J. H. Williams, J. S. Wang and T. D. Phillips, Aflatoxin and PAH exposure biomarkers in a U.S. population with a high incidence of hepatocellular carcinoma, *Sci. Total Environ.*, 2010, **408**(23), 6027–6031.
37. Y. Y. Gong, K. Cardwell, A. Hounsa, S. Egal, P. C. Turner, A. J. Hall and C. P. Wild, Dietary aflatoxin exposure and impaired growth in young children from Benin and Togo, West Africa: Cross sectional study, *Br. Med. J.*, 2002, **325,** 20–21.
38. Y. Y. Gong, A. Hounsa, S. Egal, P. C. Turner, A. E. Sutcliffe, A. J. Hall, K. Cardwell and C. P. Wild, Post-weaning exposure to aflatoxin results in impaired child growth: A longitudinal study in Benin, West Africa, *Environ. Health Persp.*, 2004 **112**, 1334–1338.
39. F. M. Shuaib, P. E. Jolly, J. E. Ehiri, Y. Jiang, W. O. Ellis, J. K. Stiles, N. J. Yatich, E. Funkhouser, S. D. Person, C. Wilson and J. H. Williams, Association between anemia and aflatoxin B1 biomarker levels among pregnant women in Kumasi, Ghana, *Am. J. Trop. Med. Hyg.*, 2010, **83**(5), 1077–1083.
40. P. C. Turner, A. Sylla, S. Y. Kuang, C. L. Marchant, M. S. Diallo, A. J. Hall, J. D. Groopman and C. P. Wild, Absence of *TP53* codon 249 mutations in Guinean infants with high aflatoxin exposure, *Cancer Epidemiol. Biomarkers Prev.*, 2005, **14**(8), 2053–2055.
41. P. C. Turner, A. C. Collinson, Y. B. Cheung, Y. Y. Gong, A. J. Hall, A. M. Prentice and C. P. Wild, Aflatoxin exposure *in utero* causes growth faltering in Gambian infants, *Int. J. Epidemiol.*, 2007, **36**, 1119–1125.
42. A. C. Collinson, P. C. Turner, A. J. Hall, C. P. Wild and A. M. Prentice, Aflatoxin-albumin biomarkers in Gambian infants, unpublished observations.
43. P. C. Turner, S. E. Moore, A. J. Hall, A. M. Prentice and C. P. Wild, Modification of immune function through exposure to dietary aflatoxin in Gambian children, *Environ. Health Persp.*, **111**(2), 217–220.
44. P. C. Turner, M. Mendy, H. Whittle, M. Fortuin, A. J. Hall and C. P. Wild, Hepatitis B infection and aflatoxin biomarker levels in Gambian children, *Trop. Med. Internat. Health*, **5**(12), 837–841.
45. S. J. Allen, C. P. Wild, J. G. Wheeler, E. M. Riley, R. Montesano, S. Bennett, H. C. Whittle and A. J. Hall, Aflatoxin exposure, malaria and hepatitis B infection in rural Gambian children, *Trans. Roy. Soc. Trop. Med. Hyg.*, 1992, **86,** 426–430.
46. H. Ahsan, L. Y. Wang, C. J. Chen, W. Y. Tsai and R. M. Santella, Variability in aflatoxin-albumin adduct levels and effects of hepatitis B and C virus infection and glutathione S-transferase M1 and T1 genotype, *Environ. Health Persp.*, 2001, **109**(8), 833–837.

47. J. D. Groopman, C. P. Wild, J. Hasler, C. Junshi, G. N. Wogan and T. W. Kensler, Molecular epidemiology of aflatoxin exposures: validation of aflatoxin-N7-guanine levels in urine as a biomarker in experimental rat models and humans, *Environ. Health Persp.*, 1993, **99**, 107–113.
48. J. Q. Zhu, L. S. Zhang, X. Hu, Y. Xiao, J. S. Chen, Y. C. Xu, J. Fremy and F. S. Chu, Correlation of dietary aflatoxin B1 levels with excretion of aflatoxin M1 in human urine, *Cancer Res.*, 1987, **47**(7), 1848–1852.
49. A. Zarba, C. P. Wild, A. J. Hall, R. Montesano, G. J. Hudson and J. D. Groopman, Aflatoxin M1 in human breast milk from The Gambia, West Africa, quantified by combined monoclonal antibody immunoaffinity chromatography and HPLC, *Carcinogenesis*, 1992, **13**(5), 891–894.
50. W. P. Blount, Turkey 'X' disease, *J. Br. Turkey Fed.*, 1961, **9**, 52–54.
51. C. M Lancaster, F. P. Jenkins and P. Philip, Toxicity associated with certain samples of groundnuts, *Nature*, 1961, **192**, 1095–1096.
52. T. Asao, G. Buchi, M. M. Abdel-Kader, S. B. Chang, E. L. Wick and G. N. Wogan, Aflatoxins B and G. *J. Am. Chem. Soc.*, 1963, **85**, 1706–1707.
53. J. M. Cullen and P. M. Newberne, Acute hepatotoxicity of aflatoxins, In *The Toxicology of Aflatoxins: Human Health, Veterinary and Agricultural Significance*, ed. D. L.Eaton and J. D. Groopman, Academic Press, 1994, pp. 3–26.
54. G. Ramjee, P. Berjak, M. Adhikari and M. F. Dutton, Aflatoxins and kwashiorkor in Durban, South Africa, *Ann. Trop. Paediat.*, 1992, **12**(3), 241–247.
55. F. E. Jonsyn-Ellis, Seasonal variation in exposure frequency and concentration levels of aflatoxins and ochratoxins in urine samples of boys and girls, *Mycopathologia*, 2001, **152**(1), 35–40.
56. Y. M. Abdulrazzaq, N. Osman, Z. M. Yousif and O. Trad, Morbidity in neonates of mothers who have ingested aflatoxins, *Ann. Trop. Paediat.*, 2004, **24**(2), 145–151.
57. Y. M. Abdulrazzaq, N. Osman and A. Ibrahim, Fetal exposure to aflatoxins in the United Arab Emirates, *Ann. Trop. Paediat.*, 2002, **22**(1), 3–9.
58. F. E. Jonsyn, S. M. Maxwell and R. G. Hendrickse, Human fetal exposure to ochratoxin and aflatoxins, *Ann. Trop. Paediat.*, 1995, **15**(1), 3–9.
59. S. M. Maxwell, J. B. Familusi, O. Sodeinde, M. C. Chan and R. G. Hendrickse, Detection of naphthols and aflatoxins in Nigerian cord blood, *Ann. Trop. Paediat.*, 1994, **14**(1), 3–5.
60. C. P. Wild, F. N. Rasheed, M. F. Jawla, A. J. Hall, L. A. Jansen and R. Montesano, In utero exposure to aflatoxin in West Africa, *Lancet*, 1991, **337**, 1602.
61. L. L. Hsieh and T. T. Hsieh, Detection of aflatoxin B1-DNA adducts in human placenta and cord blood, *Cancer Res.*, 1993, **53**(6), 1278–1280.
62. Y. Dersjant-Li, M. W. A. Verstegen and W. J. J. Gerrits, The impact of low concentrations of aflatoxin, deoxynivalenol or fumonisin in diets on growing pigs and poultry, *Nutr. Res. Rev.*, 2003, **16**, 223–239.
63. R. G. Hendrickse, Kwashiorkor and aflatoxins, *J. Pediat. Gastroenterol. Nutr.*, 1988, **7**(5), 633–636.

64. A. N. Tchana, P. F. Moundipa and F. M. Tchouanguep, Aflatoxin contamination in food and body fluids in relation to malnutrition and cancer status in Cameroon, *Int. J. Environ. Res. Public Health*, 2010, **7**(1), 178–188.
65. S. Gratz, Q. K. Wu, H. El-Nezami, R. O. Juvonen, H. Mykkänen and P. C. Turner, Alteration of intestinal transport, metabolism and toxicity of aflatoxin B1 by *Lactobacillus rhamnosus* strain GG in vitro in Caco-2 cells, *Appl. Environ. Microbiol.*, 2007, **73,** 3958–3964.
66. Y. Y. Gong, P. C. Turner, A. J. Hall and C. P. Wild, Aflatoxin exposure and impaired child growth in West Africa: an unexplored international public health burden? in *Mycotoxins Detection Methods, Management, Public Health and Agricultural Trade*, ed. J. Leslie, R. Bandyopadhyay and A. Visconti, Cromwell Press, 2008, pp. 53–66.
67. J. H. Williams, T. D. Phillips, P. E. Jolly, J. K. Stiles, C. M. Jolly and D. Aggarwal, Human aflatoxicosis in developing countries: a review of toxicology, exposure, potential health consequences, and interventions, *Am. J. Clin. Nutr.* **80**, 1106–1122.
68. B. A. Rotter, D. B. Prelusky and J. J. Pestka, Toxicology of deoxynivalenol (vomitoxin), *J. Toxicol. Environ. Health*, 1996, **48**, 1–34.
69. Y. Luo, T. Yoshizawa and T. Katayama, Comparative study on the natural occurrence of Fusarium mycotoxins (trichothecenes and zearalenone) in corn and wheat from high-risk and low-risk areas for human esophageal cancer in China, *Appl. Environ. Microbiol.*, 1990, **56**, 3723–3726.
70. R. V. Bhat, Y. Ramakrishna, S. R. Beedu and K. L. Munshi, Outbreak of trichothecene mycotoxicosis associated with consumption of mold-damaged wheat products in Kashmir Valley, India, *Lancet*, 1989, **1**, 35–37.
71. Y. Ramakrishna, R. V. Bhat and V. Ravindranath, Production of deoxynivalenol by Fusarium isolates from samples of wheat associated with a human mycotoxicosis outbreak and from sorghum cultivars, *Appl. Environ. Microbiol.*, 1989, **55**, 2619–2620.
72. C. Cowger, J. Patton-Ozkurt, G. Brown-Guedira and L. Perugini, Post-anthesis moisture increased Fusarium Head blight and deoxynivalenol levels in North Carolina winter wheat, *Phytopathology*, 2009, **99**, 320–327.
73. SCOOP Report, *Collection of Occurrence Data of Fusarium Toxins in Food and Assessment of Dietary Intake by the Population of Eu Member States*, European Commission, Brussels, 2003.
74. F. Q. Li, Y. W. Li, X. Y. Luo and T. Yoshizawa, Fusarium toxins in wheat from an area in Henan Province, Pr China, with a previous human red mould intoxication episode, *Food Addit. Contam.*, 2002, **19**, 163–167.
75. F. Q. Li, X. Y. Luo and T. Yoshizawa, Mycotoxins (trichothecenes, zearalenone and fumonisins) in cereals associated with human red-mold intoxications stored since 1989 and 1991 in China, *Nat. Toxins*, 1999, **7**, 93–97.
76. X. Luo, Food poisoning caused by Fusarium toxins, *Proceedings of Second Asian Conference in Food Safety*, 1994, pp. 129–136.

77. J. J. Pestka and A. T. Smolinski, Deoxynivalenol: toxicology and potential effects on humans, *J. Toxicol. Env. Health-Pt b-Crit. Rev.*, 2005, **8**, 39–69.
78. Centres for Disease Control (CDC), Outbreaks of gastrointestinal illness of unknown etiology associated with eating burritos – United States, October 1997–October 1998. *CDC MMWR,* 1999, **48**(10), 210–213, available at: http://www.cdc.gov/mmwr/preview/mmwrhtml/00056731.htm (accessed Feb 01, 2011).
79. P. C. Turner, V. J. Burley, J. A. Rothwell, K. L. M. White, J. E. Cade and C. P. Wild, Deoxynivalenol: Rationale for development and application of a urinary biomarker, *Food Add. Contam.*, 2008 **25**(7), 864–871.
80. P. C. Turner, Deoxynivalenol and nivalenol occurrence and exposure assessment, *World Mycotoxin J.*, **3**(4), 315–321.
81. D. B. Prelusky, R. G. Gerdes, K. L. Underhill, B. A. Rotter, P. Y. Jui and H. L. Trenholm, Effects of low-level dietary deoxynivalenol on haematological and clinical parameters of the pig, *Nat. Toxins*, 1994, **2**, 97–104.
82. L. G. Young, L. McGirr, V. E. Valli, J. H. Lumsden and A. Lun, Vomitoxin in corn fed to young-pigs, *J. Anim. Sci.*, 1983, **57**, 655–664.
83. D. M. Hughes, M. J. Gahl, C. H. Graham and S. L. Grieb, Overt signs of toxicity to dogs and cats of dietary deoxynivalenol, *J. Anim. Sci.*, 1999, **77**, 693–700.
84. F. Iverson, C. Armstrong, E. Nera, J. Truelove, S. Fernie, P. Scott, R. Stapley, S. Hayward and S. Gunner, Chronic feeding study of deoxynivalenol in B6c3f1 male and female mice, *Teratogen. Carcinogen. Mutagen.*, 1995, **15**, 283–306.
85. G. S. Bondy and J. J. Pestka, Immunomodulation by fungal toxins, *J. Toxicol. Env. Health-Pt b-Crit. Rev.*, 2000, **3**, 109–143.
86. C. J. Amuzie, J. Shinozuka and J. J. Pestka, Induction of suppressors of cytokine signaling by the trichothecene deoxynivalenol in the mouse, *Toxicol. Sci.*, 2009, **111**, 277–287.
87. J. J. Pestka and C. J. Amuzie, Tissue distribution and proinflammatory cytokine gene expression following acute oral exposure to deoxynivalenol: comparison of weanling and adult mice, *Food Chem. Toxicol.*, 2008, **46**, 2826–2831.
88. P. Pinton, F. Accensi, E. Beauchamp, A. M. Cossalter, P. Callu, F. Grosjean and I. P. Oswald, Ingestion of deoxynivalenol (Don) contaminated feed alters the pig vaccinal immune responses, *Toxicol. Lett.*, 2008, **177**, 215–222.
89. H. R. Zhou, D. Yan and J. J. Pestka, Induction of cytokine gene expression in mice after repeated and subchronic oral exposure to vomitoxin (deoxynivalenol): differential toxin-induced hyporesponsiveness and recovery, *Toxicol. Appl. Pharm.*, 1998, **151**, 347–358.
90. S. Danicke, K. P. Brussow, H. Valenta, K. H. Ueberschar, U. Tiemann and M. Schollenberger, On the effects of graded levels of Fusarium toxin contaminated wheat in diets for gilts on feed intake, growth performance

and metabolism of deoxynivalenol and zearalenone, *Mol. Nutr. Food Res.*, 2005, **49**, 932–943.
91. P. He, L. G. Young and C. Forsberg, Microbial transformation of deoxynivalenol (vomitoxin), *Appl. Environ. Microbiol.*, 1992, **58**, 3857–3863.
92. K. Seeling, S. Danicke, H. Valenta, H. P. Van Egmond, R. C. Schothorst, A. A. Jekel, P. Lebzien, M. Schollenberger, E. Razzazi-Fazeli and G. Flachowsky, Effects of Fusarium toxin-contaminated wheat and feed intake level on the biotransformation and carry-over of deoxynivalenol in dairy cows, *Food Addit. Contam.*, 2006, **23**, 1008–1020.
93. T. Yoshizawa, L. M. Cote, S. P. Swanson and W. B. Buck, Confirmation of Dom-1, a deepoxidation metabolite of deoxynivalenol, in biological-fluids of lactating cows, *Agric. Biol. Chem.*, 1986, **50**, 227–229.
94. J. C. Young, T. Zhou, H. Yu, H. H. Zhu and J. H. Gong, Degradation of trichothecene mycotoxins by chicken intestinal microbes, *Food Chem. Toxicol.*, 2007, **45**, 136–143.
95. G. S. Eriksen and H. Pettersson, Lack of de-epoxidation of Type B trichothecenes in incubates with human faeces, *Food Addit. Contam.*, 2003, **20**, 579–582.
96. F. A. Meky, P. C. Turner, A. E. Ashcroft, J. D. Miller, Y. L. Qiao, M. J. Roth and C. P. Wild, Development of a urinary biomarker of human exposure to deoxynivalenol, *Food Chem. Toxicol.*, 2003, **41**, 265–273.
97. P. C. Turner, V. J. Burley, J. A. Rothwell, K. L. M. White, J. E. Cade and C. P. Wild, Dietary wheat reduction decreases the level of urinary deoxynivalenol in UK adults, *J. Exp. Sci. Environ. Epidemiol.*, 2008, **18**, 392–399.
98. P. C. Turner, J. A. Rothwell, K. L. M. White, Y. Gong, J. E. Cade and C. P. Wild, Urinary deoxynivalenol is correlated with cereal intake in individuals from the United Kingdom, *Environ. Health Persp.*, 2008, **116**, 21–25.
99. P. C. Turner, K. L. M. White, V. Burley, R. Hopton, A. Rajendram, J. Fisher, J. E. Cade and C. P. Wild, A comparison of deoxynivalenol intake and urinary deoxynivalenol in UK adults, *Biomarkers*, 2010 **15**(6), 553–562.
100. P. C. Turner, R. P. Hopton, Y. Lecluse, K. L. M. White, J. Fisher and P. Lebailly, Determinants of urinary deoxynivalenol and de-epoxy deoxynivalenol in male farmers from Normandy, France, *J. Agric. Food Chem.*, 2010, **58**, 5206–5212.
101. S. Piekkola, P. C. Turner, E. Savchenko, S. Ezzat, I. Kasem, T. Poussa, Y. Y. Y. Chung, H. Mykkänen, M. Abdel-Hamid M and H. El-Nezami, Characterization of aflatoxin and deoxynivalenol exposure among pregnant Egyptian women, Submitted to *Food Chem. Toxicol.*
102. S. J. Hepworth, L. J. Hardie, L. K. Fraser, V. J. Burley, R. S. Mijal, C. P. Wild, R. Azad, P. A. McKinney and P. C. Turner, Deoxynivalenol exposure assessment in a cohort of pregnant women from Bradford, UK, *Food Add. Contam.*, in press.

103. P. C. Turner, R. Hopton, K. L. M. White, J. Fisher, J. E. Cade and C. P. Wild, Assessment of deoxynivalenol metabolite profiles in UK adults, *Food Chem. Toxicol.*, 2011, **49**, 132–135.
104. SCF, Opinion of the Scientific Committee on Food on Fusarium toxins. *Part 6: Group Evaluation of T-2 toxin, HT-2 toxin, Nivalenol and Deoxynivalenol: 2002SCF/CS/CNTM/MYC/27*, http://europa.eu.int/comm/food/fs/sc/scf/out123_en.pdf.
105. K. Hochstenbach, D. M. van Leeuwen, H. Gmuender, S. B. Stolevik, U. C. Nygaard, M. Lovik, B. Granum, E. Namork, J. H. M. van Delft and H. van Loveren, Transcriptomic profile indicative of immunotoxic exposure: in vitro studies in peripheral blood mononuclear cells, *Toxicol. Sci.*, 2010, **118**, 19–30.
106. A. Nogueira da Costa, R. S. Mijal, J. N. Keen, J. B. C. Findlay and C. P. Wild, Proteomic analysis of the effects of the immunomodulatory mycotoxin deoxynivalenol, *Proteomics*, 2010, in press.
107. A. Nogueira da Costa, J. N. Keen, C. P. Wild and J. B. C. Findlay, An analysis of the phosphoproteome of immune cell lines exposed to the immunomodulatory mycotoxin deoxynivalenol, *Biochim. Biophy. Acta–Prot. Proteom.*, 2010, Submitted.
108. A. M. Osman, J. L. A. Pennings, M. Blokland, A. Peijnenburg and H. van Loveren, Protein expression profiling of mouse thymoma cells upon exposure to the trichothecene deoxynivalenol (Don): Implications for its mechanism of action, *J. Immunotoxicol.*, 2010, **7**, 147–156.
109. R. P. Hopton, E. Turner, V. J. Burley, P. C. Turner and J. Fisher, Urine metabolite analysis as a function of deoxynivalenol exposure: an NMR-based metabolomics investigation, *Food Add. Contam. A*, 2010, **27**, 255–261.
110. W. F. O. Marasas, Discovery, occurrence of the fumonisins, a historical perspective, *Environ. Health Persp.*, 2001, **109**(S2), 239–243.
111. K. A. Voss, G. W. Smith and W. M. Haschek, Fumonisins: Toxicokinetics, mechanism of action and toxicity, *Anim. Feed Sci. Technol.*, 2007, **137**, 299–325.
112. W. C. Gelderblom, N. P. Kriek, W. F. Marasas and P. G. Thiel, Toxicity and carcinogenicity of the *Fusarium moniliforme* metabolite, fumonisin B1, in rats, *Carcinogenesis*, 1991, **12**, 1247–1251.
113. W. C. A. Gelderblom, S. Riedel, H. M. Burger, S. Abel and W. F. O. Marasas, Carcinogenesis by the fumonisins, mechanisms, risk analyses, and implications, in *Food Contaminants, Mycotoxins and Food Allergens*, ed. D. P. Siantar, M. W. Trucksess, P. M. Scott, E. M. Herman, A. C. S. Symposium Series 1001, Washington, DC, 2008, pp. 80–95.
114. P. C. Howard, R. M. Eppley, M. E. Stack, A. Warbritton, K. A. Voss, R. J. Lorentzen, R. M. Kovach and T. J. Bucci, Fumonisin B1 carcinogenicity in a two-year feeding study using F344 rats and B6C3F1 mice, *Health Perspect.*, 2001, **109**(S2), 277–282.
115. International Agency for Research on Cancer (IARC), Some traditional herbal medicines, some mycotoxins, naphthalene and styrene, in *Monographs on the Evaluation of Carcinogenic Risks to Humans*,

International Agency for Research on Cancer Press, Lyon (France), 2002, pp. 301–366.
116. L. Xu, Q. Cai, L. Tang, S. Wang, X. Hu, J. Su, G. Sun, J.S. Wang, Evaluation of fumonisin biomarkers in a cross-sectional study with two high-risk populations in China, *Food Add. Contam. A*, 2010, **27**, 1161–1619.
117. W. F. O. Marasas, R. T. Riley, K. A. Hendricks, V. L. Stevens, T. W. Sadler, J. Gelineau-van Waes, S. A. Missmer, J. Cabrera, O. Torres, W. C. Gelderblom, J. Allegood, C. Martinez, J. Maddox, J. D. Miller, L. Starr, M. C. Sullards, A. V. Roman, K. A. Voss, E. Wang and A. H. Merrill, Jr., Fumonisins disrupt sphingolipid metabolism, folate transport, and neural tube development in embryo culture and in vivo, a potential risk factor for human neural tube defects among populations consuming fumonisin contaminated maize, *J. Nutr.*, 2004, **134,** 711–716.
118. S. A. Missmer, L. Suarez, M. Felkner, E. Wang, A. H. Merrill Jr., K. J. Rothman and K. A. Hendricks, Exposure to fumonisins and the occurrence of neural tube defects along the Texas-Mexico border, *Environ. Health Persp.*, 2006, **114**, 237–241.
119. J. B. Gelineau-van Waes, K. A. Voss, V. L. Stevens, M. C. Speer and R. T. Riley, Maternal fumonisin exposure as a risk factor for neural tube defects, *Adv. Food Nutr. Res.*, 2009, **56**, 145–181.
120. M. Bolger, R. D. Coker, M. DiNovi, D. Gaylor, W. C. Gelderblom, M. Olsen, N. Paster, R. T. Riley, G. S. Shephard and G. J. A. Speijers, Fumonisins, in *Safety Evaluation of Certain Mycotoxins in Food*, Food Additives Series No. 47, FAO Food and Nutrition Paper No. 47, Prepared for the 56th Meeting of the Joint FAO/WHO Expert Committee on Food Additives (JECFA), World Health Organization (WHO), Geneva (Switzerland), 2001, pp. 103–279.
121. G. S. Shephard, N. L. Leggott, S. Stockenström, N. I. M. Somdyala and W. F. O. Marasas. Preparation of South African maize porridge: effect on fumonisin mycotoxin levels, *S. Afr. J. Sci.*, 2002, **98,** 393–396.
122. G. S. Shephard, W. F. O. Marasas, H. M. Burger, N. I. Somdyala, J. P. Rheeder, L. Van der Westhuizen, P. Gatyeni and D. J. Van Schalkwyk, Exposure assessment for fumonisins in the former Transkei region of South Africa, *Food Add. Contam.*, 2007, **24**, 621–629.
123. H. M. Burger, M. J. Lombard, G. S. Shephard, J. R. Rheeder, L. Van der Westhuizen and W. C. A. Gelderblom, Dietary fumonisin exposure in a rural population of South Africa, *Food Chem. Toxicol.*, 2010, **48**, 2103–2108.
124. W. T. Shier, The fumonisin paradox: A review of research on oral bioavailability of fumonisin B1, a mycotoxin produced by *Fusarium moniliforme*, *J. Toxin Rev.*, 2000, **19**, 161–187.
125. G. S. Shephard, P. G. Thiel and E. W. Sydenham, Initial studies on the toxicokinetics of fumonisin B1 in rats, *Food Chem. Toxicol.*, 1992, **30**, 277–279.

126. D. B. Prelusky, J. D. Miller and H. L. Trenholm, Disposition of 14C-derived residues in tissues of pigs fed radiolabelled fumonisin B1, *Food Add. Contam.*, 1996, **13,** 155–162.
127. W. P. Norred, R. D. Plattner and W. J. Chamberlain, Distribution and excretion of 14C-fumonisin B1 in male Sprague-Dawley rats, *Nat. Toxins*, 1993, **1**, 341–346.
128. G. S. Shephard, P. G. Thiel, E. W. Sydenham, J. F. Alberts and W. C. Gelderblom, Fate of a single dose of the 14C-labelled mycotoxin, fumonisin B1, in rats, *Toxicon*, 1992, **30**, 768–770.
129. G. S. Shephard, P. G. Thiel, E. W. Sydenham, J. F. Alberts and M. E. Cawood. Distribution and excretion of a single dose of the mycotoxin fumonisin B1 in a non-human primate, *Toxicon*, 1994, **32**, 735–741.
130. G. S. Shephard, P. G. Thiel, E. W. Sydenham and J. F. Alberts, Biliary excretion of the mycotoxin fumonisin B1 in rats, *Food Chem. Toxicol.*, 1994, **32**, 489–491.
131. G. S. Shephard, P. G. Thiel, E. W. Sydenham and M. E. Savard, Fate of a single dose of 14C-labelled fumonisin B1 in vervet monkeys, *Nat. Toxins*, 1995, **3**, 145–150.
132. G. S. Shephard, P. G. Thiel, E. W. Sydenham and P. W. Snijman, Toxicokinetics of the mycotoxin fumonisin B2 in rats, *Food Chem. Toxicol.*, 1995, **33**, 591–595.
133. G. S. Shephard and P. W. Snijman, Elimination and excretion of a single dose of the mycotoxin fumonisin B2 in a non-human primate, *Food Chem. Toxicol.*, 1999, **37**, 111–116.
134. P. Dilkin, G. Direito, M. M. Simas, C. A. Mallmann and B. Correa, Toxicokinetics and toxicological effects of single oral dose of fumonisin B1 containing *Fusarium verticillioides* culture material in weaned piglets, *Chem. Biol. Interact.*, 2010, **185**, 157–162.
135. A. H. Merrill Jr., M. C. Sullards, E. Wang, K. A. Voss and R. T. Riley, Sphingolipid metabolism, roles in signal transduction and disruption by fumonisins, *Environ. Health Perspect.*, 2001, **109**(Suppl 2), 283–289.
136. R. T. Riley, N. H. An, J. L. Showker, H. S. Yoo, W. P. Norred, W. J. Chamberlain, E. Wang, A. H. Merrill Jr., G. Motelin, V. R. Beasley and W. M. Haschek, Alteration of tissue and serum sphinganine to sphingosine ratio, an early biomarker for exposure to fumonisin-containing feeds in pigs, *Toxicol. Appl. Pharm.*, 1993, **118,** 105–112.
137. G. S. Shephard, L. Van der Westhuizen and V. Sewram, Biomarkers of exposure to fumonisin mycotoxins: A review, *Food Add. Contam.*, 2007, **24**, 1196–1201.
138. Q. Cai, L. Tang and J. S. Wang, Validation of fumonisin biomarkers in F344 rats, *Toxicol. Appl. Pharmacol.*, 2007, **225**, 28–39.
139. M. Castegnaro, L. Garren, D. Galendo, W. C. Gelderblom, P. Chelule, M. F. Dutton and C. P. Wild, Analytical method for the determination of sphinganine and sphingosine in serum as a potential biomarker

for fumonisin exposure, *J. Chromatog. B Biomed. Sci. App.*, 1998, **720,** 15–24.
140. L. Van der Westhuizen, G. S. Shephard and D. J. Van Schalkwyk, The effect of repeated gavage doses of fumonisin B1 on the sphinganine and sphingosine levels in vervet monkeys, *Toxicon*, 2001, **39,** 969–972.
141. P. N. Nikiéma, L. Worrillow, A. S. Traore, C. P. Wild and C. P. Turner, Fumonisin exposure and the sphinganine/sphingosine ratio in urine, serum and buccal cells in adults from Burkina Faso, West Africa, *World Mycotoxin J.*, 2008, **1,** 483–491.
142. M. Qiu and X. Liu, Determination of sphinganine, sphingosine and sphinganine/ sphingosine ratio in urine of humans exposed to dietary fumonisin B1, *Food Add. Contam.*, 2001, **18**, 263–269.
143. M. Solfrizzo, S. N. Chulze, C. Mallmann, A. Visconti, A. De Girolamo, F. Rojo and A. Torres, Comparison of urinary sphingolipids in human populations with high and low maize consumption as a possible biomarker of fumonisin dietary exposure, *Food Add. Contam.*, 2004, **21**, 1090–1095.
144. L. J. Silva, C. M. Lino and A. Pena, Sphinganine-sphingosine ratio in urine from two Portuguese populations as biomarker to fumonisins exposure, *Toxicon*, 2009, **54**, 390–398.
145. L. Van der Westhuizen, N. L. Brown, W. F. O. Marasas, S. Swanevelder and G. S. Shephard, Sphinganine/sphingosine ratio in plasma and urine as a possible biomarker for fumonisin exposure in humans in rural areas of Africa, *Food Chem. Toxicol.*, 1999, **37**, 1153–1158.
146. L. Van der Westhuizen, G. S. Shephard, J. P. Rheeder, N. I. M. Somdyala and W. F. O. Marasas, Sphingoid base levels in humans consuming fumonisin contaminated maize from rural areas in the former Transkei, South Africa, A cross sectional study. *Food Add. Contam. A*, 2008, **11**, 1385–1391.
147. L. Van der Westhuizen, G. S. Shephard, J. P. Rheeder and H. M. Burger, Individual fumonisin exposure and sphingoid base levels in rural populations consuming maize in South Africa, *Food Chem. Toxicol.*, 2010, **48,** 1698–1703.
148. P. H. Shetty and R. V. Bhat, Sensitive method for the detection of fumonisin B1 in human urine, *J. Chromatogr. B*, 1998, **705**, 171–173.
149. Y. Y. Gong, L. Torres-Sanchez, L. Lopez-Carrillo, J. H. Peng, A. E. Sutcliffe, K. L. White, H. U. Humpf, P. C. Turner and C. P. Wild, Association between tortilla consumption and human urinary fumonisin B1 levels in a Mexican population, *Cancer Epidemiol. Biomarkers Prev.*, 2008, **17**, 688–694.
150. L. Van der Westhuizen, Y. Y. Gong, G. S. Shephard, H. M. Burger, J. P. Rheeder, W. C. Gelderblom and C. P. Wild, Optimising sorting and washing of home-grown maize to reduce fumonisin contamination under laboratory-controlled conditions, *Food Control*, 2010, in press, DOI, 10.1016/j.foodcont.2010.09.009.

151. L. Van der Westhuizen, G. S. Shephard, H. M. Burger, J. P. Rheeder, W. C. Gelderblom, C. P. Wild and Y. Y. Gong, Implementation of simple intervention methods to reduce fumonisin exposure in a subsistence maize farming community of South Africa, *Food Add. Contam.*, 2010, **27,** 1582–1588.
152. L. Van der Westhuizen, G. S. Shephard, H. M. Burger, J. P. Rheeder, W. C. Gelderblom, C. P. Wild and Y. Y. Gong, Fumonisin B1 as a urinary biomarker of exposure in a maize intervention study among South African subsistence farmers, *Cancer Epidemiol. Biomarkers Prev.*, 2011, in press.
153. V. Sewram, N. Mshicileli, G. S. Shephard, W. F. O. Marasas, Fumonisin mycotoxins in human hair, *Biomarkers*, 2003, **8**, 110–118.
154. P. K. Chelule, N. Gqaleni, M. F. Dutton and A. A. Chuturgoon, Exposure of rural and urban populations in KwaZulu Natal, South Africa, to fumonisin B_1 in maize, *Environ. Health Perspect.,* 2001, **109**, 253–256.

CHAPTER 7

Biological Measures and the Psychosocial Working Environment

ÅSE MARIE HANSEN,* ANNE HELENE GARDE AND MARIE AARREBO JENSEN

The National Research Centre for the Working Environment, Lersø Parkallé 105, DK-2100 Copenhagen Ø, Denmark

7.1 Introduction

There is a pressing need for investigations into the biological pathways linking stress and health.[1] Endocrine factors have become increasingly relevant for the understanding of the adaptation processes and in the pathogenesis of chronic diseases caused by an adverse psychosocial working environment. An intricate network of hormones and hormone-like activities is implicated in the stress response. Exposure to an adverse psychosocial working environment initiates a number of physiological reactions, regulated by hormones.[2] The locus coeruleus–noradrenaline/autonomic (sympathetic) nervous system[3] and the hypothalamus–pituitary–adrenal (HPA) axis are the major physiological stress response systems in the body.[3–9] Further, the characterization of an individual's HPA axis activity, reactivity pattern to psychosocial stress, and inter- and intra-individual variability appear to be of major interest.[10,11]

Figure 7.1 presents a model of how the psychosocial working environment may lead to disease. The Job Demand–Control model identifies two crucial job aspects: job demand and job control.[12] Job demand refers to the workload, and has been operationalized mainly in terms of time pressure and role conflicts. Job control refers to the person's ability to control his or her work activities.

Issues in Toxicology No. 10
Biomarkers and Human Biomonitoring, Volume 2: Selected Biomarkers of Current Interest
Edited by Lisbeth E. Knudsen and Domenico Franco Merlo

Published by the Royal Society of Chemistry, www.rsc.org

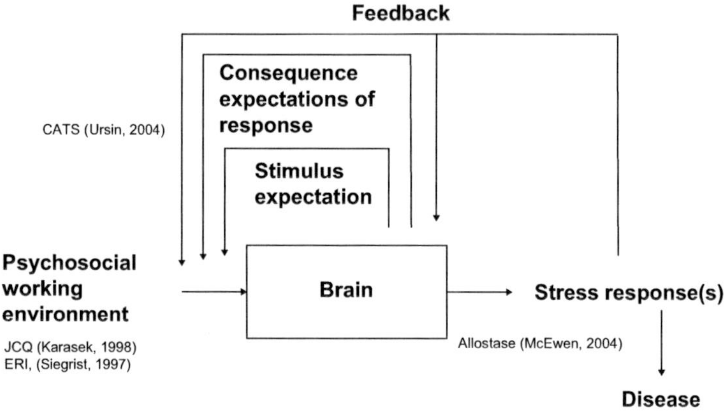

Figure 7.1 A model of the association between the psychosocial working environment and disease.

The job content questionnaire (JCQ) has been used to characterize the psychosocial working environment according to the Job Demand–Control model.[13] The Effort–Reward Imbalance (ERI) model is a model of occupational stress, focusing on a negative trade-off between experienced "costs" and "gains" at work. In this model, a high ratio of effort spent relative to rewards received, in terms of money, esteem, job security, and career opportunities, elicits sustained stress responses and ill health.[14] According to the cognitive activation theory of stress (CATS) a stress response is a general alarm in a homeostatic system, producing general and non-specific neurophysiological activation from one level of arousal to more arousal.[15] The stress response occurs whenever there is something missing, for instance a homeostatic imbalance has occurred, or there is a threat to homeostasis and the life of the organism. The stress response, therefore, is an essential and necessary physiological response. The unpleasantness of the alarm is no health threat. However, if sustained, the response may lead to illness and disease through established pathophysiological processes (allostatic load).[16]

Stress responses may be identified as changes in physiological indicators, *e.g.* endogenous substances measurable in blood, urine or saliva. Physiological indicators are therefore potential intermediate biomarkers of effect, as defined by the World Health Organization (WHO):[17] "*Biomarker for effect: a measurable biochemical, physiological, behavioural or other alteration within an organism, that depending upon the magnitude, can be recognized as associated with an established or possible health impairment or disease*". All of the physiological indicators presented in the present chapter have established associations with health impairment or disease when deviations from "normality" are large. However, the effects observed in relation to the psychosocial working environment are often smaller than those seen in clinical settings. Therefore, a relationship to health impairment or disease still remains to be established for the relevant changes in the physiological indicators.

The aim of the present chapter is twofold. One aim is to provide the reader with insight into the present evidence for how different physiological responses may be used as potential biomarkers of the psychosocial working environment. The other aim is to address and thereby bring to awareness potential sources of variation and confounders.

7.2 Psychosocial Working Environment and Stress Response

An adverse psychosocial working environment has been associated with disease. It has been shown that there is a high risk of cardiovascular diseases (CVD) among employees performing mentally straining work,[18–20] monotonous work,[21–23] as well as high-paced and shift work.[21] Long-term stress has been shown to influence the immune system and susceptibility to infection.[24] A debilitated immune defence system may lead to cancer, infections and allergy.

A possible mechanism is through stress responses elicited in response to an adverse psychosocial working environment. One commonly suggested biological mechanism behind stress-related poor health is alteration in the activity of the HPA axis.[4] The HPA axis plays a central role in homeostatic processes and is commonly thought to reflect attempts to adjust to daily pressures and joys.[25] Cortisol is also of interest in the mounting documentation in support of the view that CVD may be influenced by psychoneuroendocrine mechanisms. Some of the strongest evidence for the involvement of a psychoneuroendocrine mechanism in CVD comes from a recent prospective study, which showed an association between the cortisol-to-testosterone ratio and the incidence of ischaemic heart disease.[26] In this study, a relatively high excretion of cortisol in relation to testosterone was found to be detrimental, and the authors concluded that the relationship seemed to be mediated through the insulin resistance syndrome. Furthermore, obesity and failure to down-regulate the normally high plasma cortisol levels in the morning were found to be positively associated with indicators of poor cardiovascular health in a cross-sectional population study.[27] In another study, low diurnal variability in salivary cortisol was suggested to be mediating the association between known risk factors and CVD, type 2 diabetes, and stroke.[25]

7.3 Physiological Indicators

The physiological indicators presented in this chapter are grouped as indicators for: (1) *catabolic processes*, (2) *anabolic processes*, (3) *other metabolic processes*, and indicators of the (4) *immune response*. In the following we present the physiological effects and possible associations with disease of the selected physiological indicators. In addition, possible effects of age, gender and other important confounders, together with reference intervals for populations in work, are presented.

One way of categorizing metabolic processes, whether at the cellular, organ or organism level is as "anabolic" or as "catabolic", which are opposites. *Catabolism* is the set of processes that breaks down molecules into smaller units and releases energy. In catabolism, large molecules such as polysaccharides, lipids, nucleic acids and proteins are broken down into smaller units such as monosaccharides, fatty acids, nucleotides and amino acids, respectively. There are many signals that control catabolism. Most of the known signals are hormones and the molecules involved in metabolism itself. *Anabolism* is the set of metabolic processes that constructs molecules from smaller units. These reactions require energy. Anabolic processes tend toward "building up" organs and tissues. These processes produce growth and differentiation of cells and increases in body size, a process that involves synthesis of complex molecules.[28]

7.3.1 Psyiological Indicators for Catabolic Processes

Adrenaline and noradrenaline: Adrenaline and noradrenaline (catecholamines) are the traditional "fight-or-flight" hormones released by the adrenal glands in response to stress. They are part of the sympathetic nervous system. They may be used as physiological indicators of activation of the central nervous system (CNS). Chronically increased levels of catecholamines (primarily adrenaline and noradrenaline) in blood correlate with increased blood pressure and the risk of ischaemic heart disease.[22,29] Concentrations of adrenaline and noradrenaline vary with age and gender, with highest levels in middle-aged men.[30,31] Reference intervals in urine have previously been established for healthy individuals ($n = 119$) at work to be from the limit of detection (LOD) to 2.86 μmol adrenaline/mol creatinine and 3.6–29.1 μmol noradrenaline/mol creatinine in morning samples (05.45–07.15 h); from LOD to 9.84 μmol adrenaline/mol creatinine and 11.0–54.1 μmol noradrenalin/mol creatinine in afternoon samples (15.30–18.30 h); and from LOD to 13.20 μmol adrenaline/mol creatinine and 7.3–49.2 μmol noradrenaline/mol creatinine in evening samples (21.45–23.45 h).[32]

Cortisol: The HPA axis plays a central role in homeostatic processes and is commonly thought to reflect our attempts to adjust to our daily pressures and joys.[25] The principal marker for activation of the HPA axis in field studies and much experimental stress research is cortisol, although several hormones are involved [*e.g.* corticotropin-releasing hormone (CRH) and adrenocorticotropic hormone (ACTH)]. The immediate physiological effects of increased cortisol concentrations, for example increased blood pressure, suppression of inflammation and precipitation of insulin resistance, are well documented.[33–35] The long-term effects of minor deviation in the excretion of cortisol are, however, less well documented although it is known that prolonged and elevated concentrations of cortisol may lead to the redistribution of body fat, characterized by trunk obesity, hypertension, and type 2 diabetes, as seen in Cushing's disease.[27,33] Further, neuroendocrine changes such as lower cortisol levels may be associated with the underlying pathology of post-traumatic stress disorder (PTSD).[36] In addition, patients with major depression exhibit higher concentrations of cortisol during the recovery period than non-depressed individuals.[37]

In humans, perceived stress and other factors activate the CNS, with the release of cortisol-releasing hormone (CRH) from the hypothalamus, which stimulates release of adrenocorticotrophic hormone (ACTH) from the anterior pituitary, and thereby cortisol from the adrenal cortex. In turn, elevations in cortisol levels typically inhibit the HPA system *via* negative feedback mechanisms in the hippocampus. This is part of our naturally occurring regulation system.

No difference in serum cortisol between age groups has been reported.[30] Reference intervals in urine have been established previously for healthy individuals ($n = 119$) at work to be from 2.3–52.8 μmol cortisol/mol creatinine in morning samples (05.45–07.15 h); from LOD to 42.4 μmol cortisol/mol creatinine in afternoon samples (15.30–18.30 h); and from LOD to 17.6 μmol cortisol/mol creatinine in evening samples (21.45–23.45).[32]

Cholesterol: Cholesterol is a steroid metabolite found in the cell membranes and transported in the blood plasma of all animals. It is an essential structural component of mammalian cell membranes, where it is required to establish proper membrane permeability and fluidity. In addition, cholesterol is an important component for the manufacture of bile acids, steroid hormones, and fat-soluble vitamins. It is well known that an increased level of cholesterol over a long period may lead to arteriosclerosis.[21] Levels of cholesterol in serum have been followed over a period of 20 years for women and 10 years for men in one study with 8737 persons. Levels of cholesterol decreased 1% with age for both groups. The decrease could be due to a change in the ratio between saturated and unsaturated fat or life style changes.[38] A recent study among healthy individuals in a small Danish population, however, showed that the concentration of cholesterol was dependent on age and gender. The reference interval for total cholesterol in serum of men and women between 30 and 60 years was found to be 3.52–7.86 mmol/L.[39] This is in line with common knowledge.[40]

Haemoglobin A_{1c}: Haemoglobin A_{1c} (HbA_{1c}) reflects the serum glucose concentration of the past 4 to 5 weeks[41] and is useful for the screening of diabetes mellitus.[42] A reference interval was established for healthy subjects at work as 3.4–5.60% HbA_{1c} of total haemoglobin. The amount of HbA_{1c} increases with age, and smokers exhibit higher concentrations than non-smokers.[39]

7.3.2 Physiological Indicators for *anabolic processes*

Testosterone and oestrogens: Testosterone and oestrogens are anabolic hormones that stimulate synthesis of proteins. Testosterone and oestrogen concentrations vary with gender and age.[28] Among healthy subjects, reference intervals were established in blood for men to be 31.5–92.0 pmol/L and for women 0.30–7.70 pmol/L.[39] Oestradiol was measured in a random sample of postmenopausal women drawn from the base population in Denmark in

1981–1983. Among 832 postmenopausal women without hormone treatment a 95% confidence interval was found for oestradiol to be 7–123 pmol/L and among 210 hormone-treated postmenopausal women to be 20 2351 pmol/L.[43]

Dehydroepiandrosterone (DHEA) is a weak anabolic steroid and therefore expected to be related to physical stress. In the blood more than 99% of DHEA is bound to sulfate as the more stable DHEA-S. The concentration of DHEA decrease with age.[28] The reference interval of DHEA-S in serum in a working population was 1.07–11.04 μmol/L. DHEA-S decreases with age and increases with body mass index (BMI).[39]

7.3.3 Physiological Indicators for Other Metabolic Processes

Prolactin: Several factors stimulate the secretion of prolactin in serum, *e.g.* exercise and stress. The prolactin level is at its highest during sleep.[31] The reference interval in serum for men was 11–329 mIU/L and for women 9–621 mIU/L.[39]

Melatonin: Melatonin is released from the pineal gland in response to changes in light perceived through the retina.[44,45] The secretion of melatonin is closely related to light intensity, hence seasonal variation in daylight is an important confounder.[48] A peak concentration of melatonin occurs at about 2 to 3 a.m. during normal night sleeping conditions.[46] The half-life of human melatonin in blood is about 35–50 min, and it is metabolized mainly in the liver. The major metabolite in urine, 6-sulfatoxymelatonin, reflects approximately the blood concentration.[44] Exposure to even short periods of visible light of sufficient intensity during the normal nocturnal circadian phase will immediately decrease the production and release of melatonin in a dose–response related manner.[45,47] It has been suggested that decreased levels of melatonin, which may be predicted as a consequence of shift and night work, may increase the risk of breast cancer.[47,48]

7.3.4 Physiological Indicators of the Immune Response and Other Parameters

The immune system is extremely sensitive to a number of hormones, such as cortisol and testosterone.[28] *Acute phase proteins* are proteins of which the concentration in plasma increases by more than 25% after activation of the immune system,[38] hence *fibrinogen* has been included in the present chapter.

7.4 Physiological Indicators and the Psychosocial Working Environment

Physiological systems activated in response to the psychosocial working environment may both restore and harm the body.[7,49] Measurement of biological indicators in blood, urine or saliva may add important information on the effects of the total work exposure of the individual. Stress responses promote adaptation, but prolonged over- or under-activation of physiological systems leads over time to wear-and-tear on the body (allostatic load). One of the first studies on the psychosocial working environment and physiological indicators was on overtime work. It was found that levels of urinary adrenaline measured during the day and in the evening were consistently elevated during the overtime period and 4 weeks after the overtime period ended among office workers.[50] Stressful and poorly organized working environments as well as deficiencies in leadership may facilitate work-related bullying either directly or by creating a work climate in which bullying can flourish.[51–55]

Physiological indicators have been used to assess the effects of exposure to the working environment such as job demands, job control, *etc.*).[56–61] See Table 7.1

Table 7.1 Examples of studies on physiological indicators and the psychosocial working environment. DHEA-S, dehydroepiandrosterone; HDL, high-density lipoprotein; LDL, low-density lipoprotein.

Physiological indicator	*Matrix*	*Psychosocial working environment evaluated*	*References*
Cortisol	Blood, urine and saliva	Job demands/control, leadership, lack of support, low job satisfaction, effort-reward imbalance, shift work, monotonous work, bullying	(62,62,63, 63–76) (77–87)
Cholesterol (HDL, LDL and total)	Blood	Effort–reward imbalance	(63,66,88–105)
DHEA-S	Blood	Job demands, job control, leadership, and organizational changes	(63,66,72,75, 83,106,107)
Oestrogen	Blood	Job demands, job control, social support, and organizational changes	(63,67,83)
Fibrinogen	Blood	Job demands, job control, social support, organizational changes and effort–reward	(75,88,92, 108–112)
Glycated haemoglobin (HbA_{1c})	Blood	Job demands/control, lack of support, low job satisfaction, shift work, occupational changes	(41,63,66,75, 113–115)
Melatonin	Saliva	Shift work	(116,117)
Noradrenaline and adrenaline	Urine	Monotony/high work pace, shift work	(66–70,75,77, 118–124)
Prolactin	Blood	Job demands/control	(62,63,66,67, 72–76,78, 83–85,91)
6-Sulfatoxymelatonin	Urine	Shift work	(85,125)
Testosterone	Blood and saliva	Job demands/control	(62,66,67,73, 74,76,78,84, 85,116,117, 126,127)

In the present chapter the psychosocial working environment includes job strain, which covers job demands (cognitive and quantitative work demands), and job control, denoted by the demand–control model.[12] The effort–reward imbalance model was developed to include reward, which may be a major factor in the psychosocial working environment,[128] and other psychosocial factors such as leadership (predictability, reward, role conflicts), social support (from colleagues and supervisors) and job satisfaction/job dissatisfaction.

Catecholamines: Catecholamines in blood have only been used in a few studies on the psychosocial working environment. They were positively associated with monotony/high work pace[77,120] and having shift work.[121–124] Urinary catecholamines have been used in workplace studies, but not in population-based studies. A reason for this could be that collecting urine for the measurement of catecholamines requires a tight sampling procedure, where the samples have to be frozen within 24 h and the tube to have acid added or be coated with acid[129] in order to prevent the catecholamines from degrading. Further, urinary catecholamines exhibit a distinct diurnal variation and have therefore often been sampled several times a day[130] in order to analyse associations with exposure levels. Altogether, collection of urine for measurement of catecholamines requires close contact with the participants in order to reduce variability. As a consequence it is difficult and costly to study the associations with urinary catecholamines in studies with a high number of participants because very high demands for sampling logistics must be set. Among studies of the association between urinary catecholamines and the psychosocial working environment only one study exceeded 100 participants[66] and no significant associations were found. This may be a result of imprecise sample collection.

Cortisol: Among studies addressing associations between cortisol in serum or urine and the psychosocial working environment no consistent results were found: some showed no association,[62,63,66–74] others positive association,[75–83] and others a negative association.[78,84,85] Cortisol was found to be positively associated with organizational changes,[75,76,83] whereas no associations were observed with monotonous work,[70] shift work[73] or effort–reward imbalance.[71,88] Positive associations have been observed between concentrations of cortisol and poor leadership,[81] lack of support,[80–82] and low job satisfaction.[82] Salivary cortisol has been measured among bullied workers and was found to be lower in this population.[131–133] A recent study distinguished between frequent and occasional bullying and found that only the frequent bullying was associated with low salivary cortisol.[134]

The studies have included both urinary cortisol[66,70,75,77,80,82] and serum cortisol.[62,63,65,67,71–74,76,78,79,81,83–85] One reason for the discrepancies between studies may be that cortisol exhibits a distinct diurnal variation[130] and the sampling schedules in the studies were not comparable. Thus, irrespective of the medium, the sampling schedule may be critical for the comparison of results from different studies. Alternatively, biological variation, *e.g.* in terms of diurnal variation, seasonal variation and age-related variation, should be considered when planning a surveillance programme by identifying specific sampling schedules and calendar periods.[130]

Cholesterol: Cholesterol has been used widely in both population-based and workplace studies as well as longitudinal and cross-sectional studies. One of the first studies on physiological indicators and an adverse psychosocial working environment was performed among tax accountants, in whom long-term exposure to high workload was related to increased levels of cholesterol and an increased tendency for blood to coagulate.[105] Later, cholesterol showed the most robust results in population-based studies. Among the population-based studies that showed a positive association, three out of four were longitudinal. Of the 11 studies that showed no association between cholesterol and the psychosocial working environment nine were cross-sectional and two longitudinal.[135] Hence the study design may have an impact on the measurement of physiological responses to the psychosocial working environment.

HbA_{1c}: Studies on HbA_{1c} in serum consistently show positive associations with the psychosocial working environment.[41,63,66,75,113–115] This is true for both longitudinal (three studies) and cross-sectional (four studies) designs. A recent review found that the result was clearest in relation to the job-demand model, where all studies on HbA_{1c} reported positive associations to both job demands[63,66,113,114] and job control.[63,66,92,113] Positive associations of concentrations of HbA_{1c} with lack of support[63,66,113] and low job satisfaction[41] have been observed. Concentrations of HbA_{1c} were also positively associated with shift work[114] and occupational changes.[75] Hence HbA_{1c} is regarded as a robust indicator for the effect of the psychosocial working environment, independent of the type of study. This is in line with the common understanding in models of how the psychosocial working environment may affect the physiological systems, *i.e.* to perform in a job it is necessary to mobilize the catabolic systems.[7]

Testosterone: Anabolic indicators are expected to decrease with an adverse psychosocial working environment. Concentrations of testosterone were negatively associated with job demands[136] and job control.[66,67,78] With respect to testosterone, a negative association was observed with shift work,[73,85] but not with organizational changes.[76] Serum testosterone decreased[66,67,73,78,84,85] when the psychosocial working environment was perceived to be adverse. These results covered both longitudinal and cross-sectional studies, and this finding is consistent with common knowledge about the anabolic system.[7] The anabolic system is down-regulated and the catabolic system is up-regulated when exposed to stress.

DHEA-S: The picture is more blurred with respect to oestrogen and DHEA-S, which showed both positive and negative associations with the psychosocial working environment.[63,66,67,72,75,83,106,107]

Prolactin: Concentrations of prolactin were negatively associated with job demands[62,91,136] and positively associated with job control.[62,66,67,72,78,91] Only one study was longitudinal and here a positive association with job control was found.[67]

Melatonin: Night work is associated with lower concentrations of 6-sulfatoxy-melatonin in urine[125,137–139] in most studies. Altogether, night workers tend to have less melatonin than day workers, which may be one of the hypothesized mechanisms associated with the increased risk of breast cancer which has been found consistently in recent observational studies of night- and shift workers.[140,141]

Fibrinogen: All population-based cross-sectional studies using fibrinogen in blood showed a positive association with an adverse psychosocial working environment.[108–110,112] Concentrations of fibrinogen were positively associated with demands in all population-based studies,[108–110,112] but not in workplace studies.[92,111] Hence fibrinogen appears to be a robust and sensitive physiological indicator for the psychosocial working environment.

7.4.1 Summary

A recent review[135] stated that concentrations of fibrinogen and catabolic indicators were increased, and the concentrations of anabolic indicators were decreased when the psychosocial working environment was perceived to be poor, such as with high job demands, low job control, *etc.* The metabolic indicators and indicators of effects on the immune system were both increased and decreased in relation to an adverse psychosocial working environment. The most consistent results in the review by Hansen *et al.*[135] were observed for HbA_{1c}, testosterone and fibrinogen. Thus, HbA_{1c}, testosterone and fibrinogen in serum may be relevant as biomarkers for the effect of the psychosocial working environment.

7.5 How may the Results be Biased from the Sampling Procedure?

When collecting samples by self-monitoring in field studies there is less control on adherence to the protocol compared to studies on hospitalized patients. It is therefore essential to make protocols as simple as possible and to estimate the consequences of lack of adherence. Several examples are given below.

7.5.1 Biological Variation

In occupational health studies, the study groups most often comprise healthy subjects performing their work. Sampling is therefore often planned in the most practical way, *e.g.* sampling of blood in the morning at the work site just after the work starts or by self-monitoring over 24 h, *e.g.* by sampling of saliva. Changes in concentrations of hormones when performing routine work are a mixture of responses to physiological and psychosocial exposure and normal biological variations, *e.g.* diurnal variation, seasonal variation, within- and between-subject variation, *etc.*[142–146] Within- and between-subject variation may be related to gender, age and life-style factors such as smoking, physical exercise and alcohol intake.[147] To assess effects of exposure to the psychosocial working environment it is crucial to distinguish between the normal biological variations and effects related to the working environment.

Reference intervals for clinical purposes are often established for hospitalized reference subjects living under the same conditions, but not suffering from the same disease, as the patients.[148] Hence, the published reference intervals are not necessarily suitable for use in occupational medicine. The International Federation of Clinical Chemistry and Laboratory Medicine (IFCC) recommends that the parametric confidence intervals of fractiles of reference intervals should be based on an approximation to the normal distribution after transformation of reference intervals.[149] For estimating reference intervals as recommended by the IFCC, samples from more than 120 subjects are recommended.[150] If the component in question is influenced by, for example, age, BMI, gender or smoking it can be costly and difficult to recruit group sizes of that order of magnitude. Hansen *et al.* presented a method to estimate reference intervals for smaller sample sizes.[39] This method also makes it possible to test for other biological variations. Figure 7.2 shows one example of biological variation, in HbA_{1c} in blood, among the active working population (37 women and 84 men aged 30–60 years). It shows that HbA_{1c} increases with both smoking and age.

Serum androgens decrease with age.[151–160] Bernini *et al.* (2001) reported that the decrease is independent of menopause.[154] Hansen *et al.* found that the age-related decrease for concentrations of serum DHEA-S was 27% per 10 years, which corresponds to a 70-year-old subject having a concentration of serum DHEA-S which is 2.6 times lower than that of a 30-year-old subject.[160] In contrast, concentrations of DHEA-S increase with BMI (body mass index).

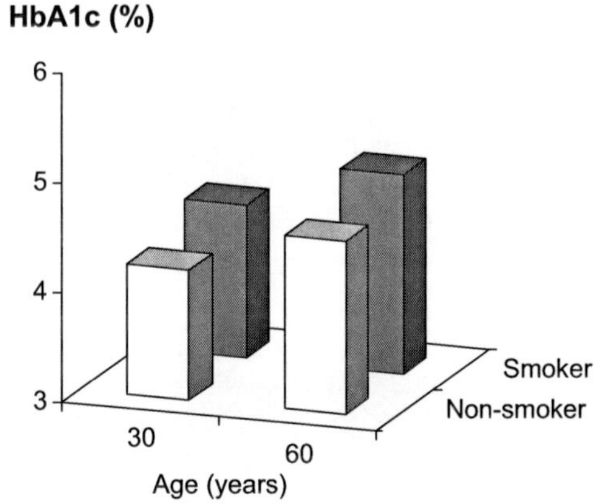

Figure 7.2 HbA_{1c} in blood among smoking and non-smoking men and women at work aged 30 and 60 years.

The effect of BMI on concentrations of S-DHEA-S is 4% per kg/m^2. Blood HbA_{1c} was shown to increase by 4% per 10 years of age.[160,161] Concentrations of free testosterone were negatively correlated with age and positively correlated with BMI in women, but not in men.[162]

7.5.2 Seasonal Variations

Seasonal variations are known to influence many aspects of human physiology and behaviour, for example hormone concentrations, heart rate variability and mood.[130,163–166] As this may affect the interpretation of study results, such knowledge is important. As an example, seasonal variations have been described for total cholesterol, DHEA-S, HbA1c, prolactin and free testosterone in blood from individuals at work. The concentrations of HbA_{1c} and DHEA-S exhibited a cyclic variation with a period of 6 months.[163] The difference between the 3 months with the highest and that with the lowest concentrations was 12% for total cholesterol, 9% for DHEA-S, 7% for HbA_{1c}, 44% for prolactin, and 31% for free testosterone based on the predicted values. No seasonal effects were observed for IgA.[163]

7.5.3 Sampling Saliva for Salivary Cortisol: Cotton *versus* Polyester

The Salivette® is a widely used method for collection of saliva. It involves the use of a tampon, which is placed in the mouth until soaked with saliva. After the tampon has been replaced in its container the saliva may be extracted by centrifugation. Tampons based on different materials may be purchased, and the choice of material has been found to influence the measurement of, for example, steroid components.[167,168] In a recent study it was found that cotton tampons reduced the measured concentration of cortisol to 62% of that found when using polyester tampons.[168] Further, the cotton tampons retained more saliva than polyester. Recovery of saliva from polyester tampons was 95.8($\pm$1.1)% and from cotton tampons 54.7($\pm$2.3)%. This is most relevant, because a huge problem in sampling saliva is problems with small volumes of saliva. The use of additives to raise the production of saliva has been suggested. A widely used additive is citric acid.[169] However, citric acid interferes with many immunoassays, and it must be stressed not to use the Salivette® with citric acid in order to increase saliva production when using immunoassays without testing the effect first.

7.5.4 Compliance with Sample Collection

When using self-monitoring, it is crucial to gain knowledge about compliance with the sampling procedures. An important factor when sampling saliva is compliance with the time of sampling.[170] Some studies have used electronic devices to track when participants actually accessed the cotton swab, or tampon. In one study it was observed that 74% of the participants accessed the tampon according to the study protocol, whereas 26% failed to access the tampon at the proper time for at least one out of six samples. Of this latter group of non-compliants, 55% failed to take the second morning sample correctly after 30 minutes. Participants who were not informed that their sampling was being tracked were significantly less compliant than informed participants.[171] Another study that examined participant adherence found that 71% of participants, who were unaware that they were being monitored, followed the protocol correctly. Their self-reported compliance was, however, 93%. Among the persons who were aware of being monitored, the objective compliance was 90%, consistent with the self-reported compliance of 93%.[172] In both studies, the non-adherent participants had significantly lower morning cortisol values than the adherent participants.

7.5.5 Creatinine or Volume of Urine?

When measuring biomarkers in urine, total volume (and time), *e.g.* 24-hour samples or concentrations of creatinine, are both accepted methods of standardization for diuresis. Both types of standardization contribute to the uncertainty in the final result. It has been found that the uncertainty associated with creatinine standardization (19–35%) was higher than the uncertainty related to volume standardization (up to 10%, when not correcting for deviations from 24 h) for 24 h urine samples.[173] Even so, standardization with creatinine is often used owing to a considerable increase in convenience for the participants when collecting small volumes rather than complete 24 h samples.

7.5.6 Storage

In research projects, samples are often required to be stored for long periods of time either because of the protocol of the project or because of lack of funding for analysis. A study on long-term storage found no effects on cortisol concentrations after storage of saliva at 5 °C for up to 3 months or at –20 °C and –80 °C for up to 1 year. In contrast, concentrations of cortisol were found to decrease by 9.2% [95% confidence interval (CI): 3.8–14.3%] per month in samples stored at room temperature.[174] Repeated freezing and thawing of samples up to four times before analysis did not affect the measured concentrations of cortisol.[174,175] To improve the stability of urine samples for catecholamine determinations, the samples have to be collected in acidified tubes. A simple way of doing this may be to precoat 10 mL polystyrene tubes with 200 mg citric acid.[176] The acidified urine samples are stable for approximately 2 months at –20 °C (see Table 7.2).

Kley *et al.* examined whether plasma samples may be used for analysis of steroids after long periods of storage; cortisol, testosterone, oestrone and oestradiol were re-measured in samples which had been analysed 1.3–10.8 years earlier.[180] The method for the measurement of these steroids was unchanged

Table 7.2 Examples of biological indicators in saliva, blood and urine.

Example of biological indicators in:	*Collection*	*Duration of storage (Ref)*[a]
Saliva		
Cortisol	Salivette® (no additive)	Approximately 1 year after centrifugation (178)
Melatonin	Salivette® (no additive)	No data available
Testosterone	Straw or tubes no additive	> 28 days (179)
Blood:		
Total cholesterol	Serum	2% decrease per year (180)
HDL-cholesterol	Serum	2% decrease per year (180)
HbA_{1c}	Whole blood (EDTA)	1–2 years[b]
Free testosterone	Serum	2 months (manufacturer of assay)/10 years(181)
Oestradiol	Serum	10 years (181)
Fibrinogen	Plasma/serum	1.5 years (182)
Prolactin	Serum	1.5 years (183)
Urine		
Adrenaline	Tube coated with citric acid	Less than 2 months
Noradrenaline	Tube coated with citric acid	Less than 2 months
Cortisol	Plain tube/ bottle	No data
Creatinine	Plain tube/bottle	>3 months (184)

[a]The longer the samples are to be stored the more the variance will increase.
[b]Laboratory observations, not published.

over this period. The results demonstrated that at a temperature of –25 °C steroids remained stable. Only cortisol and testosterone concentrations showed a small, insignificant decrease (6–9%) after 3 to 4 years of storage. These differences are well within the range of the precision of the method (inter-assay variation), which over a period of 11 years was 9.4%, 8.0%, 10.0% and 9.5% for cortisol, testosterone, oestrone and oestradiol, respectively. It was concluded that steroid hormones in human plasma were stable, and that they might be analysed even after more than 10 years of storage at –25 °C.[180]

7.5.7 Summary

Based on the present knowledge, a few recommendations have been given to help ensure a practical protocol and increase the quality of occupational field studies.[184] Governing thoughts behind the recommendations were that they should: (a) provide as few restraints and as little inconvenience for the participants as possible, (b) serve to help reduce unnecessary variability in the study design, and (c) provide suggestions for dealing with variability in cases where influences are unavoidable:

(i) To account for the diurnal variation, the time of sampling should be carefully registered and, if necessary, included in the statistical analysis, or samples should be obtained at the same time of day.

(ii) To prevent bias being introduced in longitudinal studies by seasonal variations, samples should be collected at the same time of year.
(iii) To avoid bias from variation caused by using different laboratory techniques, it is recommended that the same, or comparable, laboratory techniques are used.
(iv) To avoid bias from the material in the tampon for collecting saliva specimens, it is recommended that the same material be used throughout the study or studies.
(v) To avoid bias from storage, it is recommended that saliva samples for measurement of cortisol are stored at –20 °C or below, because the changes in concentrations at this temperature will be negligible for 1 year. If there is a need to store for longer periods, an industrial freezer is recommended (–80 °C).
(vi) It is recommended to collect and use information regarding smoking, alcohol consumption, medication, *etc.*

References

1. B. M. Kudielka and S. Wüst, Human models in acute and chronic stress: assessing determinants of individual hypothalamus-pituitary-adrenal axis activity and reactivity, *Stress*, 2010, **13**, 1–14.
2. J. P. Henry, Biological basis of the stress response, *Integr. Physiol. Behav. Sci.*, 1992, **27**, 66–83.
3. G. P. Chrousos and P. W. Gold, The concepts of stress and stress system disorders. Overview of physical and behavioral homeostasis, *J. Am. Med. Assoc.*, 1992, **267**, 1244–1252.
4. B. S. McEwen and T. Seeman, Protective and damaging effects of mediators of stress. Elaborating and testing the concepts of allostasis and allostatic load, *Ann. N. Y. Acad. Sci.*, 1999, **896**, 30–47.
5. C. Tsigos and G. P. Chrousos, Hypothalamic-pituitary-adrenal axis, neuroendocrine factors and stress, *J. Psychosom. Res.*, 2002, **53**, 865–871.
6. P. W. Gold, J. Licinio, M. L. Wong and G. P. Chrousos, Corticotropin releasing homone in the pathophysiology of melancholic and atypical depression and in the mechanism of action of antidepressant drugs, *Ann. N. Y. Acad. Sci.*, 1995, **771**, 716–729.
7. B. S. McEwen, Protective and damaging effects of stress mediators, *New Eng. J. Med.*, 1998, **338**, 171–179.
8. C. Heim, U. Ehlert and D. H. Hellhammer, The potential role of hypocortisolism in the pathophysiology of stress-related bodily disorders, *Psychoneuroendocrinology*, 2000, **25**, 1–35.
9. C. L. Raison and A. H. Miller, When not enough is too much: the role of insufficient glucocorticoid signaling in the pathophysiology of stress-related disorders, *Am. J. Psychiatry*, 2003, **160**, 1554–1565.
10. J. W. Mason, A review of psychoendocrine research on the pituitary-thyroid system, *Psychosom. Med.*, 1968, **30**, 666–681.

11. D. H. Hellhammer, S. Wüst and B. M. Kudielka, Salivary cortisol as a biomarker in stress research, *Psychoneuroendocrinology*, 2009, **34**, 163–171.
12. R. Karasek and T. Theorell, Job structure and physiological response, in *Healthy work. Stress, Productivity, and the Reconstruction of Working Life*, ed. R. Karasek and T. Theorell, Basic Books, Inc., New York, 1990, pp. 103–109.
13. R. Karasek, C. Brisson, N. Kawakami, I. Houtman, P. Bongers and B. Amick, The Job Content Questionnaire (JCQ): an instrument for internationally comparative assessments of psychosocial job characteristics, *J. Occup. Health Psychol.*, 1998, **3**, 322–355.
14. J. Siegrist, D. Starke, T. Chandola, I. Godin, M. Marmot, I. Niedhammer and R. Peter, The measurement of effort-reward imbalance at work: European comparisons, *Soc. Sci. Med.*, 2004, **58**, 1483–1499.
15. H. Ursin and H. R. Eriksen, The cognitive activation theory of stress, *Psychoneuroendocrinology*, 2004, **29**, 567–592.
16. B. S. McEwen and J. C. Wingfield, The concept of allostasis in biology and biomedicine, *Horm. Behav.*, 2003, **43**, 2–15.
17. World Health Organization (WHO) *Biomarkers – Biomarkers and Risk Assessment: Concepts and Principles*, WHO, Geneva, 1993.
18. A. LaCroix and S. Haynes, Occupational exposure to high demand/low control work and coronary heart disease incidence in the Framingham cohort, *Am. J. Epidemiol.*, 1984, **120**, 481–481.
19. R. A. Karasek, D. Baker, F. Marxer, A. Ahlbom and T. Theorell, Job decision latitude, job demands, and cardiovascular disease: A prospective study of swedish men, *Am. J. Public Health*, 1981, **71**(7), 694–705.
20. C. Pieper, A. Z. LaCroix and R. A. Karasek, The relation of psychosocial dimensions of work with coronary heart disease risk factors: a meta-analysis of five united states data bases, *Am. J. Epidemiol.*, 1989, **129**(3), 483–494.
21. T. S. Kristensen, Cardiovascular diseases and the work environment. A critical review of the epidemiologic literature on nonchemical factors, *Scand. J. Work Environ. Health*, 1989, **15**, 165–179.
22. N. J. Christensen, Katekolaminer og psykisk stress, *Ugeskr. Læger*, 1986, **148**, 233–237.
23. V. Putz-Anderson, G. T. Doyle and T. R. Hales, Ergonomic analysis to characterize task constraint and repetitiveness as risk factors for musculoskeletal disorders in telecommunication office work, *Scand. J. Work Environ. Health*, 1992, **18**(Suppl 2), 123–126.
24. S. Cohen and G. M. Williamson, Stress and infectious disease in humans, *Psychol. Bull.*, 1991, **109**(1), 5–24.
25. R. Rosmond and P. Björntorp, Occupational status, cortisol secretory pattern, and visceral obesity in middle-aged men, *Obes. Res.*, 2000, **8**, 445–450.
26. G. D. Smith, Y. Ben Shlomo, A. Beswick, J. Yarnell, S. Lightman and P. Elwood, Cortisol, testosterone, and coronary heart disease: prospective evidence from the Caerphilly study, *Circulation*, 2005, **112**, 332–340.

27. B. R. Walker, S. Soderberg, B. Lindahl and T. Olsson, Independent effects of obesity and cortisol in predicting cardiovascular risk factors in men and women, *J. Intern. Med.*, 2000, **247**, 198–204.
28. E.-E. Baulieu and P. A. Kelly, *Hormones: from Molecules to Disease*, Herman, Paris, France, 1990, p. 697.
29. T. Matsukawa, E. Gotoh, S. Uneda, E. Miyajima, H. Shionoiri, O. Tochikubo and M. Ishii, Augmented sympathetic nerve activity in response to stressors in young borderline hypertensive men, *Acta Physiol. Scand.*, 1991, **141**, 157–165.
30. A. Sudo, Evaluation of workload in middle-aged steel workers by measuring urinary excretion of catecholamines and cortisol, *Jpn. J. Ind. Health*, 1991, **33**(6), 475–484.
31. T. F. Mejman, H. P. G. Mulders, M. A. J. Kompier and M. Dormolen, Individual differences in adrenalin/noradrenalin reactivity and self-perceived health status, *Z. Gesamte. Hyg.*, 1990, **36**(8), 413–414.
32. Å. M. Hansen, A. H. Garde, J. M. Christensen, N. H. Eller and B. Netterstrøm, Reference intervals and variation for urinary epinephrine, norepinephrine and cortisol in healthy men and women in Denmark, *Clin. Chem. Lab. Med.*, 2001, **39**, 842–849.
33. B. R. Walker, D. I. Phillips, J. P. Noon, M. Panarelli, R. Andrew, H. V. Edwards, D. W. Holton, J. R. Seckl, D. J. Webb and G. C. Watt, Increased glucocorticoid activity in men with cardiovascular risk factors, *Hypertension*, 1998, **31**, 891–895.
34. R. C. Andrews, O. Herlihy, D. E. Livingstone, R. Andrew and B. R. Walker, Abnormal cortisol metabolism and tissue sensitivity to cortisol in patients with glucose intolerance, *J. Clin. Endocrinol. Metab.*, 2002, **87**, 5587–5593.
35. P. Björntorp and R. Rosmond, Neuroendocrine abnormalities in visceral obesity, *Int. J. Obes.*, 2000, **24**(Suppl 2), S80–S85.
36. R. Yehuda, B. Kahana, K. Binder-Brynes, S. M. Southwick, J. W. Mason and E. L. Giller, Low urinary cortisol excretion in Holocaust survivors with post-traummatic stress disorder, *Am. J. Psychiatry*, 1995, **152**, 982–986.
37. H. M. Burke, M. C. Davis, C. Otte and D. C. Mohr, Depression and cortisol responses to psychological stress: A meta-analysis, *Psychoneuroendocrinology*, 2005, **30**, 846–856.
38. A. Sjøl, K. Grunnet and M. Schroll, Secular trends in serum cholesterol, high density lipoproteins and triglycerides 1964-1987, *Int. J. Epidemiol.*, 1991, **20(1)**, 105–113.
39. Å. M. Hansen, A. H. Garde and N. H. Eller, Estimation of individual reference intervals in small sample sizes, *Int. J. Hyg. Environ. Health*, 2007, **210**, 471–478.
40. I. Lorentzen, G. Bendixen and N. E. Hansen, *Medical Compendium [Medicinsk kompendium]*, Nyt Nordisk Forlag Arnold Busk, Copenhagen, 1999, p. 2908.
41. N. Kawakami, S. Araki, T. Hayashi and T. Masumoto, Relationship between perceived job-stress and glycosylated hemoglobin in white-collar workers, *Ind. Health*, 1989, **27**, 149–154.

42. W. F. Ganong, *Review of Medical Physiology*, Lange Medical Publications, Los Altos, California, 1983, p. 654.
43. N. H. Rod, Å. M. Hansen, J. Nielsen, P. Schnohr and M. Grønbæk, Low risk factor profile, estrogen levels, and breast cancer risk among postmenopausal women, *Int. J. Cancer*, 2009, **1254**, 1935–1940.
44. C. A. Czeisler, J. F. Duffy, T. L. Shanahan, E. N. Brown, J. F. Mitchell, D. W. Rimmer, J. M. Ronda, E. J. Silva, J. S. Allan, J. S. Emens, D.-J. Dijk and R. E. Kronauer, Stability, precision, and near-24-hour period of the human circadian pacemaker, *Science*, 1999, **284**, 2177–2181.
45. A. Brzezinski, Melatonin in humans, *N. Engl. J. Med.*, 1997, **336**, 186–195.
46. D. Dawson and N. Encel, Melatonin and sleep in humans, *J. Pineal Res.*, 1993, **15**, 1–12.
47. R. G. Stevens and M. S. Rea, Light in the built environment: potential role of circadian disruption in endocrine disruption and breast cancer, *Cancer Causes Control*, 2001, **12**, 279–287.
48. R. G. Stevens and B. W. Wilson, *The Melatonin Hypothesis – Breast Cancer and Use of Electric Power*, Battelle Press, Ohio, 1997, p. 760.
49. H. Ursin and H. R. Eriksen, The cognitive activation theory of stress, *Psychoneuroendocrinology.*, 2004, **29**, 567–592.
50. M. Frankenhaeuser and G. Johansson, Stress at work: psychobiological and psychosocial aspects, *Int. Rev. Appl. Psychol.*, 1986, **35**, 287–299.
51. N. A. Bowling and T. A. Beehr, Workplace harassment from the victim's perspective: a theoretical model and meta-analysis, *J Appl. Psychol*, 2006, **91**, 998–1012.
52. M. Agervold and E. G. Mikkelsen, Relationships between bullying, psychosocial work environment and individual stress reactions, *Work & Stress*, 2004, **18**, 336–351.
53. S. Einarsen, B. I. Raknes and S. B. Matthiesen, Bullying and harassment at work and their relationship to work environment quality-an exploratory study, *Eur. Work Organizat. Psychol.*, 1994, **4**, 381–401.
54. H. Leymann, The content and development of mobbing at work, *Eur. J. Work Organizat. Psychol.*, 1996, **5**, 165–184.
55. L. J. Hauge, A. Skogstad and S. Einarsen, Relationships between stressful work environments and bullying: Results of a large representative study, *Work & Stress*, 2007, **21**, 220–242.
56. B. S. McEwen and E. Stellar, Stress and the individual. Mechanisms leading to disease, *Arch. Intern. Med.*, 1993, **157**, 2093–2101.
57. T. M. Pollard, Physiological consequences of everyday psychosocial stress, *Coll. Antropol.*, 1997, **21**, 17–28.
58. T. Theorell, Possible mechanisms behind the relationship between the demand-control-support model and disorders of the locomotor system, in *Psychosocial Aspects of Musculoskeletal Disorders in Office Work*, ed. S. D. Moon and S. L. Sauter, Taylor & Francis, London, 1996, pp. 65–73.
59. J. K. Sluiter, M. H. W. Frings-Dresen, T. F. Meijman and A. J. van der Beek, Reactivity and recovery from different natures of work measured by

catecholamines and cortisol: A systematic literature overview, *Occup. Environ. Med.*, 2000, **57**, 298–315.

60. N. Hjortskov, A. H. Garde, P. Ørbæk and A. M. Hansen, Evaluation of salivary cortisol as a biomarker of self-reported mental stress in field studies, *Stress Health*, 2004, **20**, 91–98.
61. Å. M. Hansen, A. D. Larsen, R. Rugulies, A. H. Garde and L. E. Knudsen, A review of the effect of the psychosocial working environment on physiological changes in blood and urine, *Basic Clin. Pharmacol. Toxicol.*, 2009, **105**, 73–83.
62. C. G. Ohlson, M. Söderfeldt, B. Söderfeldt, I. Jones and T. Theorell, Stress markers in relation to job strain in human service organizations, *Psychother. Psychosom*, 2001, **70**, 268–275.
63. G. Grossi, A. Perski, B. Evengard, V. Blomkvist and K. Orth-Gomer, Physiological correlates of burnout among women, *J. Psychosom. Res.*, 2003, **55**, 309–316.
64. T. Theorell, A. Perski, T. Akerstedt, F. Sigala, G. Ahlberg-Hulten, J. Svensson and P. Eneroth, Changes in job strain in relation to changes in physiological state. A longitudinal study, *Scand. J. Work Environ. Health*, 1988, **14**, 189–196.
65. A. Evolahti, M. Hultcrantz and A. Collins, Women's work stress and cortisol levels: a longitudinal study of the association between the psychosocial work environment and serum cortisol, *J. Psychosom. Res.*, 2006, **61**, 645–652.
66. Å. M. Hansen, A. Kaergaard, J. H. Andersen and B. Netterstrøm, Associations between repetitive work and endocrinological indicators of stress, *Work & Stress*, 2003, **17**, 264–276.
67. M. Berg, B. B. Arnetz, S. Lidén, P. Eneroth and A. Kallner, Technostress. A psychophysiological study of employees with VDU-associated skin complaints, *J. Occup. Med.*, 1992, **34**, 698–701.
68. A. Härenstam and T. P. Theorell, Work conditions and urinary excretion of catecholamines-a study of prison staff in Sweden, *Scand. J. Work Environ. Health*, 1988, **14**, 257–264.
69. G. Vivoli, M. Bergomi, S. Rovesti, G. Carrozzi and A. Vezzosi, Biochemical and haemodynamic indicators of stress in truck drivers. Special Issue: Psychophysiological measures in transport operations, *Ergonomics*, 1993, **36**, 1089–1098.
70. K. M. Smirnov, A. A. Viru and T. A. Smirnova, Comparative analysis of urinary corticosteroid and catecholamine levels in persons with different occupational training, *Hum. Physiol.*, 1985, **11**, 351–354.
71. M. Irie, A. Tsutsumi, I. Shioji and F. Kobayashi, Effort-reward imbalance and physical health among Japanese workers in a recently downsized corporation, *Int. Arch. Occup. Environ. Health*, 2004, **77**, 409–417.
72. T. Theorell, G. Ahlberg-Hulten, M. Jodko, F. Sigala and B. de la Torre, Influence of job strain and emotion on blood pressure in female hospital personnel during workhours, *Scand. J. Work Environ. Health*, 1993, **19**, 313–318.

73. J. Axelsson, T. Åkerstedt, G. Kecklund, A. Lindqvist and R. Attefors, Hormonal changes in satisfied and dissatisfied shift workers across a shift cycle, *J. Appl. Physiol.*, 2003, **95**, 2099–2105.
74. T. Theorell, H. Leymann, M. Jodko, K. Konarski, H. E. Norbeck and P. Eneroth, 'Person under train' incidents: medical consequences for subway drivers, *Psychosom. Med.*, 1992, **54**, 480–488.
75. B. Netterstrøm and Å. M. Hansen, Outsourcing and stress: Physiological effects on bus drivers, *Stress Med.*, 2000, **16**, 149–160.
76. I. Anderzen and B. B. Arnetz, The impact of a prospective survey-based workplace intervention program on employee health, biologic stress markers, and organizational productivity, *J. Occup. Environ. Med.*, 2005, **47**, 671–682.
77. U. Lundberg, M. Granqvist, T. Hansson, M. Magnusson, and L. Wallin, Psychological and physiological stress responses during repetitive work at an assembly line, *Work Stress*, 1989, **3**(2), 143–153.
78. T. Theorell, G. Ahlberg-Hulten, F. Sigala, A. Perski and M. Soderholm, A psychosocial and biomedical comparison between men in six contrasting service occupations, *Work Stress*, 1990, **4**(1), 51–63.
79. R. M. Rose, C. D. Jenkins, M. Hurst, B. E. Kreger, J. Barrett and R. P. Hall, Endocrine activity in air traffic controllers at work. III. Relationship to physical and psychiatric morbidity, *Psychoneuroendocrinology*, 1982, **7**, 125–134.
80. P. Schnorpfeil, A. Noll, R. Schulze, U. Ehlert, K. Frey and J. E. Fischer, Allostatic load and work conditions, *Soc. Sci. Med.*, 2003, **57**, 647–656.
81. A. Härenstam and T. Theorell, Cortisol elevation and serum gamma-glutamyl transpeptidase in response to adverse job conditions: how are they interrelated? *Biol. Psychol.*, 1990, **31**, 157–171.
82. R. L. Payne, J. T. Rick, G. H. Smith and R. G. Cooper, Multiple indicators of stress in an active job – Cardiothoracic surgery, *J. Occup. Med.*, 1984, **26**(11), 805–808.
83. A. Hertting and T. Theorell, Physiological changes associated with downsizing of personnel and reorganisation in the health care sector, *Psychother. Psychosom.*, 2002, **71**, 117–122.
84. G. Grossi, T. Theorell, M. Jurisoo and S. Setterlind, Psychophysiological correlates of organizational change and threat of unemployment among police inspectors, *Integr. Physiol. Behav. Sci.*, 1999, **34**, 30–42.
85. Y. Touitou, Y. Motohashi, A. Reinberg, C. Touitou, P. Bourdeleau, A. Bogdan and A. Auzeby, Effect of shift work on the night-time secretory patterns of melatonin, prolactin, cortisol and testosterone, *Eur. J. Appl. Physiol.*, 1990, **60**, 288–292.
86. Y. Chida and A. Steptoe, Cortisol awakening response and psychosocial factors: a systematic review and meta-analysis, *Biol. Psychol.*, 2009, **80**, 265–278.
87. B. M. Kudielka, D. H. Hellhammer and S. Wüst, Why do we respond so differently? Reviewing determinants of human salivary cortisol responses to challenge, *Psychoneuroendocrinology*, 2009, **34**, 2–18.

88. T. G. Vrijkotte, L. J. van Doornen and E. J. de Geus, Work stress and metabolic and hemostatic risk factors, *Psychosom. Med.*, 1999, **61**, 796–805.
89. M. G. Kang, S. B. Koh, B. S. Cha, J. K. Park, J. M. Woo and S. J. Chang, Association between job stress on heart rate variability and metabolic syndrome in shipyard male workers, *Yonsei Med. J.*, 2004, **45**, 838–846.
90. G. E. Hedberg, K. A. Jacobsson, U. Janlert and S. Langendoen, Risk indicators of ischemic heart disease among male professional drivers in Sweden, *Scand. J. Work Environ. Health*, 1993, **19**, 326–333.
91. C. T. Su, Association between job strain status and cardiovascular risk in a population of Taiwanese white-collar workers, *Jpn. Circ. J.*, 2001, **65**, 509–513.
92. H. Riese, L. J. van Doornen, I. L. Houtman and E. J. de Geus, Job strain and risk indicators for cardiovascular disease in young female nurses, *Health Psychol.*, 2000, **19**, 429–440.
93. N. Kawakami, T. Haratani and S. Araki, Job strain and arterial blood pressure, serum cholesterol, and smoking as risk factors for coronary heart disease in Japan, *Int. Arch. Occup. Environ. Health*, 1998, **71**, 429–432.
94. M. Romon, M. C. Nuttens, C. Fievet, P. Pot, J. M. Bard, D. Furon and J. C. Fruchart, Increased triglyceride levels in shift workers, *Am. J. Med.*, 1992, **93**, 259–262.
95. Y. Kobayashi, T. Hirose, Y. Tada, A. Tsutsumi and N. Kawakami, Relationship between two job stress models and coronary risk factors among Japanese part-time female employees of a retail company, *J. Occup. Health*, 2005, **47**, 201–210.
96. M. Kivimäki, P. Leino-Arjas, R. Luukkonen, H. Riihimäki, J. Vahtera and J. Kirjonen, Work stress and risk of cardiovascular mortality: prospective cohort study of industrial employees, *BMJ*, 2002, **325**, 857.
97. K. Orth Gomer, Intervention on coronary risk factors by adapting a shift work schedule to biological rhythmicity, *Psychosom. Med.*, 1983, **45**(5), 407–415.
98. C. T. Su, H. J. Yang, C. F. Lin, M. C. Tsai, Y. H. Shieh and W. T. Chiu, Arterial blood pressure and blood lipids as cardiovascular risk factors and occupational stress in Taiwan, *Int. J. Cardiol.*, 2001, **81**, 181–187.
99. H. Tsuboi, A. Tatsumi, K. Yamamoto, F. Kobayashi, K. Shimoi and N. Kinae, Possible connections among job stress, depressive symptoms, lipid modulation and antioxidants, *J. Affect. Disord.*, 2006, **91**, 63–70.
100. C. Leineweber, G. Kecklund and K. Orth-Gomér, Prediction of cardio-cerebrovascular and other significant disease from disturbed sleep and work strain, *Scand. J. Work Environ. Health*, 2007, **33**, 215–222.
101. E. D. Eaker, L. M. Sullivan, M. Kelly-Hayes, R. B. D'Agostino, Sr. and E. J. Benjamin, Does job strain increase the risk for coronary heart disease or death in men and women? The Framingham Offspring Study, *Am. J. Epidemiol.*, 2004, **159**, 950–958.

102. C. Fornari, M. Ferrario, C. Menni, R. Sega, R. Facchetti and G. C. Cesana, Biological consequences of stress: conflicting findings on the association between job strain and blood pressure, *Ergonomics*, 2007, **50**, 1717–1726.
103. M. Kivimäki, M. Hintsanen, L. Keltikangas-Jarvinen, M. Elovainio, L. Pulkki-Raback, J. Vahtera, J. S. Viikari and O. T. Raitakari, Early risk factors, job strain, and atherosclerosis among men in their 30s: the Cardiovascular Risk in Young Finns Study, *Am. J. Public Health*, 2007, **97**, 450–452.
104. R. Peter, L. Alfredsson, N. Hammar, J. Siegrist, T. Theorell and P. Westerholm, High effort, low reward, and cardiovascular risk factors in employed Swedish men and women: baseline results from the WOLF Study, *J. Epidemiol. Community Health*, 1998, **52**, 540–547.
105. M. Friedman, R. H. Rosenman and V. Carroll, Changes in the serum cholesterol and blood clotting time in men subjected to cyclic variation of occupational stress, *Circulation*, 1958, **18**, 852–861.
106. T. Theorell, K. Orth Gomer and P. Eneroth, Slow-reacting immunoglobulin in relation to social support and changes in job strain: a preliminary note, *Psychosom. Med.*, 1990, **52**, 511–516.
107. T. Theorell, K. Harms Ringdahl, G. Ahlberg-Hulten and B. Westin, Psychosocial job factors and symptoms from the locomotor system – a multicausal analysis, *Scand. J. Rehab. Med.*, 1991, **23**, 165–173.
108. E. Clays, B. D. De, J. Delanghe, F. Kittel, R. L. Van and B. G. De, Associations between dimensions of job stress and biomarkers of inflammation and infection, *J. Occup. Environ. Med.*, 2005, **47**, 878–883.
109. F. Kittel, F. Leynen, M. Stam, M. Dramaix, S. P. de, R. Mak, B. G. De and M. Kornitzer, Job conditions and fibrinogen in 14226 Belgian workers: the Belstress study, *Eur. Heart J.*, 2002, **23**, 1841–1848.
110. A. Tsutsumi, T. Theorell, J. Hallqvist, C. Reuterwall and U. de Faire, Associations between job characteristics and plasma fibrinogen in a normal working population: A cross sectional analysis in referents of the SHEEP study, *J. Epidemiol. Commun. Health*, 1999, **53**, 348–354.
111. M. Ishizaki, P. Martikainen, H. Nakagawa and M. Marmot, The Japan work stress and health cohort-study group, Socioeconomic status, workplace characteristics and plasma fibrinogen level of Japanese male employees, *Scand. J. Work Environ. Health*, 2001, **27**, 287–291.
112. A. Steptoe, S. Kunz-Ebrecht, N. Owen, P. J. Feldman, A. Rumley, G. D. Lowe and M. Marmot, Influence of socioeconomic status and job control on plasma fibrinogen responses to acute mental stress, *Psychosom. Med.*, 2003, **65**, 137–144.
113. N. Kawakami, K. Akachi, H. Shimizu, T. Haratani, F. Kobayashi, M. Ishizaki, T. Hayashi, O. Fujita, Y. Aizawa, S. Miyazaki, H. Hiro, S. Hashimoto and S. Araki, Job strain, social support in the workplace, and haemoglobin A1c in Japanese men, *Occup. Environ. Med.*, 2000, **57**, 805–809.

114. G. Cesana, G. Panza, M. Ferrario, R. Zanettini, M. Arnoldi and A. Grieco, Can glucosylated hemoglobin be a job stress parameter? *J. Occup. Med.*, 1985, **27**(5), 357–360.
115. P. J. Feldman and A. Steptoe, Psychosocial and socioeconomic factors associated with glycated hemoglobin in nondiabetic middle-aged men and women, *Health Psychol.*, 2003, **22**, 398–405.
116. L. E. Carlson, T. S. Campbell, S. N. Garland and P. Grossman, Associations among salivary cortisol, melatonin, catecholamines, sleep quality and stress in women with breast cancer and healthy controls, *J. Behav. Med.*, 2007, **30**, 45–58.
117. H. R. Lieberman, G. P. Bathalon, C. M. Falco, F. M. Kramer, C. A. Morgan, III and P. Niro, Severe decrements in cognition function and mood induced by sleep loss, heat, dehydration, and undernutrition during simulated combat, *Biol. Psychiatry*, 2005, **57**, 422–429.
118. K. Rodahl, Z. Vokac, P. Fugelli and O. Vaage, Circulatory strain, estimated energy output and catecholamine excretion in Norwegian coastal fishermen, *Ergonomics*, 1974, **17**(5), 585–602.
119. Y. Fujigaki and K. Mori, Longitudinal study of work stress among information system professionals, *Int. J. Human-Computer Interact.*, 1997, **9**, 369–381.
120. M. Timio, S. Gentili and S. Pede, Free adrenaline and noradrenaline excretion related to occupational stress, *Br. Heart J.*, 1979, **42**, 471–474.
121. M. A. Levitt and G. R. Derrick, An evaluation of physiological parameters of stress in the emergency department, *Am. J. Emerg. Med.*, 1991, **9**, 217–219.
122. S. Fujiwara, S. Shinkai, Y. Kurokawa and T. Watanabe, The acute effects of experimental short-term evening and night shifts on human circadian rhythm: the oral temperature, heart rate, serum cortisol and urinary catecholamines levels, *Int. Arch. Occup. Environ. Health*, 1992, **63**, 409–418.
123. H. P. G. Mulders, T. F. Mejman, J. F. O'Hanlon and G. Mulder, Differential psychophysiological reactivity of city bus drivers, *Ergonomics*, 1982, **25**(11), 1003–1011.
124. K. Fujiwara, E. Tsukishima, S. Kasai, A. Masuchi, A. Tsutsumi, N. Kawakami, H. Miyake and R. Kishi, Urinary catecholamines and salivary cortisol on workdays and days off in relation to job strain among female health care providers, *Scand. J. Work Environ. Health*, 2004, **30**, 129–138.
125. Å. M. Hansen, A. H. Garde and J. Hansen, Diurnal urinary 6-sulfatoxymelatonin levels among healthy Danish nurses during work and leisure time, *Chronobiol. Int.*, 2006, **23**, 1203–1215.
126. R. H. Rahe, S. Karson, N. S. Jr. Howard, R. T. Rubin and R. E. Poland, Psychological and physiological assessment on American hostages freed from captivity in Iran, *Psychosom. Med.*, 1990, **52**, 1–16.
127. M. J. Muller, Salivary testosterone and simple reaction time parameters, *Neuropsychobiology*, 2000, **30**, 173–177.

128. J. Siegrist, Adverse health effects of high-effort/low-reward conditions, *J. Occup. Health Psychol.*, 1996, **1**, 27–41.
129. Å. M. Hansen, J. Kristiansen, J. L. Nielsen, K. Byrialsen and J. M. Christensen, Validation of a high performance liquid chromatography analysis for the determination of noradrenaline and adrenaline in human urine with an on-line sample purification, *Talanta*, 1999, **50**, 367–379.
130. Å. M. Hansen, A. H. Garde, L. T. Skovgaard and J. M. Christensen, Seasonal and biological variation of urinary epinephrine, norepinephrine, and cortisol in healthy women, *Clin. Chim. Acta*, 2001, **309**, 25–35.
131. Å. M. Hansen, A. Høgh, R. Persson, B. Karlson, A. H. Garde and P. Ørbæk, Bullying at work, health outcomes, and physiological stress response, *J. Psychosom. Res.*, 2006, **60**, 63–72.
132. B. M. Kudielka and S. Kern, Cortisol day profiles in victims of mobbing (bullying at the work place): preliminary results of a first psychobiological field study, *J. Psychosom. Res.*, 2004, **56**, 149–150.
133. P. Monteleone, G. Nolfe, C. Serritella, V. Milano, C. A. Di, F. Blasi, C. Petrella and M. Maj, Hypoactivity of the hypothalamo-pituitary-adrenal axis in victims of mobbing: role of the subjects' temperament and chronicity of the work-related psychological distress, *Psychother. Psychosom.*, 2009, **78**, 381–383.
134. Å. M. Hansen, A. Hogh and R. Persson, Frequency of bullying at work, physiological response, and mental health, *J. Psychosom. Res.*, 2011, **70**, 19–27.
135. Å. M. Hansen, A. D. Larsen, R. Rugulies, A. H. Garde and L. E. Knudsen, A review of the effect of the psychosocial working environment on physiological changes in blood and urine, *Basic Clin. Pharmacol. Toxicol.*, 2009, **105**, 73–83.
136. Å. M. Hansen, A. Kaergaard, J. H. Andersen and B. Netterstrøm, Associations between repetitive work and endocrinological indicators of stress, *Work Stress*, 2003, **17**, 264–276.
137. H. Yamauchi, M. Iwamoto and N. Harada, Physiological effects of shift work on hospital nurses, *J. Hum. Ergol. (Tokyo)*, 2001, **30**, 251–254.
138. V. H. H. Goh, T. Y. Y. Tong, C.-L. Lim, E. C. T. Low and L. K. H. Lee, Circadian disturbances after night-shift work onboard a naval ship, *Mil. Med.*, 2000, **165**, 101–105.
139. J. B. Burch, M. G. Yost, W. Johnson and E. Allen, Melatonin, sleep, and shift work adaptation, *J. Occup. Environ. Med.*, 2005, **47**, 893–901.
140. S. P. Megdal, C. H. Kroenke, F. Laden, E. Pukkala and E. S. Schernhammer, Night work and breast cancer risk: a systematic review and meta-analysis, *Eur. J. Cancer*, 2005, **41**, 2023–2032.
141. P. K. Verkasalo, K. Lillberg, R. G. Stevens, C. Hublin, M. Partinen, M. Koskenvuo and J. Kaprio, Sleep duration and breast cancer: a prospective cohort study, *Cancer Res.*, 2005, **65**, 9595–9600.
142. G. Y. Nicolau, E. Haus, M. Popescu, L. Sackett-Lundeen and E. Petrescu, Circadian, weekly, and seasonal variations in cardiac mortality, blood pressure, and catecholamine excretion, *Chronobiol. Int.*, 1991, **8**, 149–159.

143. G. Y. Nicolau, E. Haus, D. J. Lakatua, C. Bogdan, L. Sackett-Lundeen, M. Popescu, H. Berg, E. Petrescu and E. Roby, Circadian and circannual variations of FSH, LH, testosterone, dehydroepiandrosterone-sulfate (DHEA-S) and 17-hydroxy progesterone (17 OH-Prog) in elderly men and women, *Endocrinologie*, 1985, **23**, 223–246.
144. M. Maes, K. Mommen, D. Hendricks, D. Peeters, P. D'Hondt, R. Ranjan, F. De Meyer and S. Scharpé, Components of biological variation, including seasonality, in blood concentrations of TSH, TT3, FT4, PRL, cortisol and testosterone in healthy volunteers, *Clin. Endocrinol.*, 1997, **46**, 587–598.
145. C. G. Fraser, S. T. Cummings, S. P. Wilkinson, R. G. Neville, J. D. E. Knox, O. Ho and R. S. MacWalter, Biological variability of 26 clinical chemistry analytes in elderly people, *Clin. Chem.*, 1989, **35**, 783–786.
146. A. H. Garde, Å. M. Hansen, L. T. Skovgaard and J. M. Christensen, Seasonal and biological variation of blood concentrations of total cholesterol, dehydroepiandrosterone sulfate, hemoglobin A1C, 1gA, prolactin, and free testosterone in healthy women, *Clin Chem*, 2000, **46**, 551–559.
147. A. H. Garde, R. Persson, A. M. Hansen, K. Österberg, P. Ørbæk, F. C. Eek and B. Karlson, Effects of lifestyle factors on concentrations of salivary cortisol in healthy indviduals, *Scand. J. Clin. Lab. Invest.*, 2008, **49**, 242–250.
148. T. Kouri, V. Kairisto, A. Virtanen, E. Uusipaikka, A. Rajamaki, H. Finneman, K. Juva, T. Koivula and V. Nanto, Reference intervals developed from data for hospitalized patients: computerized method based on combination of laboratory and diagnostic data, *Clin. Chem.*, 1994, **40**, 2209–2215.
149. H. E. Solberg, Statistical treatment of reference values in laboratory medicine: Testing the goodness-of-fit of an observed distribution to the Gaussian distribution, *Scand. J. Clin. Lab. Invest.*, 1986, **46**(Suppl. 184), 125–132.
150. O. M. Poulsen, E. Holst and J. M. Christensen, Calculation and application of coverage intervals of biological reference values, *Pure Appl. Chem.*, 1997, **69**, 1601–1611.
151. Å. M. Hansen, A. H. Garde, N. Eller and B. Netterstrøm, Biological variation in blood free-testosterone, DHEA-S, HBA1C, IGA, prolactin, and total-cholesterol, in *Behavioural Medicine: Local and Global. Seventh International Congress of Behavioural Medicine*, Helsinki, Finland, Aug. 28–31 2002, p. 103.
152. E. Barrett-Connor, D. G. Von Mühlen and D. Kritz-Silverstein, Bioavailable testosterone and depressed mood in older men: the Rancho Bernardo Study, *J. Clin. Endocrinol. Metab.*, 1999, **84**, 573–577.
153. G. P. Bernini, M. Sgro', A. Moretti, G. F. Argenio, C. O. Barlascini, R. Cristofani and A. Salvetti, Endogenous androgens and carotid intimal-medial thickness in women, *J. Clin. Endocrinol. Metab.*, 1999, **84**, 2008–2012.

154. G. P. Bernini, A. Moretti, M. Sgró, G. F. Argenio, C. O. Barlascini, R. Cristofani and A. Salvetti, Influence of endogenous androgens on carotid wall in postmenopausal women, *Menopause*, 2001, **8**, 43–50.
155. L. Denti, G. Pasolini, L. Sanfelici, F. Ablondi, M. Freddi, R. Benedetti and G. Valenti, Effects of aging on dehydroepiandrosterone sulfate in relation to fasting insulin levels and body composition assessed by bioimpedance analysis, *Metabolism*, 1997, **46**, 826–832.
156. A. E. Field, G. A. Colditz, W. C. Willett, C. Longcope and J. B. McKinlay, The relation of smoking, age, relative weight, and dietary intake to serum adrenal steroids, sex hormones, and sex hormone-binding globulin in middle-aged men, *J. Clin. Endocrinol. Metab.*, 1994, **79**, 1310–1316.
157. S. M. Harman, E. J. Metter, J. D. Tobin, J. Pearson and M. R. Blackman, Longitudinal effects of aging on serum total and free testosterone levels in healthy men. Baltimore Longitudinal Study of Aging, *J. Clin. Endocrinol. Metab.*, 2001, **86**, 724–731.
158. E. A. Jankowska, E. Rogucka, M. Medras and Z. Welon, Relationships between age-related changes of sex steroids, obesity and body fat distribution among healthy Polish males, *Med. Sci. Monit.*, 2000, **6**, 1159–1164.
159. N. Thomas, H. A. Morris, F. Scopacasa, J. M. Wishart and A. G. Need, Relationships between age, dehydro-epiandrosterone sulphate and plasma glucose in healthy men, *Age Ageing*, 1999, **28**, 217–220.
160. Å. M. Hansen, A. H. Garde and N. H. Eller, Estimation of individual reference intervals in small sample sizes, *Int. J. Hyg. Environ. Health*, 2007, **210**, 471–478.
161. Y. Hashimoto, A. Futamura and M. Ikushima, Effects of aging on HbA_{1c} in a working male Japanese population, *Diabetes Care*, 1995, **18**, 1337–1340.
162. A. Vermeulen, S. Goemaere and J. M. Kaufman, Testosterone, body composition and aging, *J. Endocrinol. Invest.*, 1999, **22**, 110–116.
163. A. H. Garde, A. M. Hansen, L. T. Skovgaard and J. M. Christensen, Seasonal and biological variation of blood concentrations of total cholesterol, dehydroepiandrosterone sulfate, hemoglobin A(1c), IgA, prolactin, and free testosterone in healthy women, *Clin. Chem.*, 2000, **46**, 551–559.
164. C. D. Ferrie, T. C. Sharpe, D. A. Price and R. A. H. Surtees, Seasonal variation of glycosylated haemoglobin, *Arch. Dis. Child.*, 1987, **62**, 959–960.
165. J. Kristiansen, A. Olsen, J. H. Skotte and A. H. Garde, Reproducibility and seasonal variation of ambulatory short-term heart rate variability in healthy subjects during a self-selected rest period and during sleep, *Scand. J. Clin. Lab. Invest.*, 2009, **69**, 651–661.
166. M. Okawa, S. Shirakawa, M. Uchiyama, M. Oguri, M. Kohsaka, K. Mishima, K. Sakamoto, H. Inoue, K. Kamei and K. Takahashi, Seasonal variation of mood and behaviour in a healthy middle-aged population in Japan, *Acta Psychiatr. Scand.*, 1996, **94**, 211–216.

167. E. A. Shirtcliff, D. A. Granger, E. Schwartz and M. J. Curran, Use of salivary biomarkers in biobehavioral research: cotton-based sample collection methods can interfere with salivary immunoassay results, *Psychoneuroendocrinology*, 2001, **26**, 165–173.
168. Å. M. Hansen, A. H. Garde and R. Persson, Measurement of salivary cortisol – effects of replacing polyester with cotton and switching antibody, *Scand. J. Clin. Lab. Invest.*, 2008, **68**, 826–829.
169. F. R. Burlage, J. Pijpe, R. P. Coppes, M. E. Hemels, H. Meertens, A. Canrinus and A. Vissink, Variability of flow rate when collecting stimulated human parotid saliva, *Eur. J. Oral. Sci.*, 2005, **113**, 386–390.
170. F. C. Eek, A. H. Garde, Å. M. Hansen, R. Persson, P. Orbaek and B. Karlson, The cortisol awakening response – an exploration of intraindividual stability and negative responses, *Scand. J. Work Environ. Health*, 2006, **Suppl 2**, 15–21.
171. B. M. Kudielka, J. E. Broderick and C. Kirschbaum, Compliance with saliva sampling protocols: electronic monitoring reveals invalid cortisol daytime profiles in noncompliant subjects, *Psychosom. Med.*, 2003, **65**, 313–319.
172. C. E. Wright and A. Steptoe, Subjective socioeconomic position, gender and cortisol responses to waking in an elderly population, *Psychoneuroendocrinology*, 2005, **30**, 582–590.
173. A. H. Garde, A. M. Hansen, J. Kristiansen and L. E. Knudsen, Comparison of uncertainties related to standardization of urine samples with volume and creatinine concentration, *Ann. Occup. Hyg.*, 2004, **48**, 171–179.
174. A. H. Garde and A. M. Hansen, Long-term stability of salivary cortisol, *Scand. J. Clin. Lab. Invest.*, 2005, **65**, 433–436.
175. Å. M. Hansen, A. H. Garde, B. Arnetz, S. Knardahl, M. Kristenson, U. Lundberg, B. Malmberg, B. Netterstrøm, T. Theorell, M. T. Tuomisto and H. Ursin, *A Model Database for Salivary Cortisol in Occupational Studies*, Nordic Council of Ministers, Copenhagen, 2005, pp. 4–72.
176. A. M. Hansen, A. H. Garde, J. M. Christensen, N. H. Eller and B. Netterstrom, Reference intervals and variation for urinary epinephrine, norepinephrine and cortisol in healthy men and women in Denmark, *Clin. Chem. Lab Med.*, 2001, **39**, 842–849.
177. A. H. Garde and A. M. Hansen, Long-term stability of salivary cortisol, *Scand. J. Clin. Lab. Invest.*, 2005, **65**, 433–436.
178. F. Matsui, E. Koh, K. Yamamoto, K. Sugimoto, H. S. Sin, Y. Maeda, S. Honma and M. Namiki, Liquid chromatography-tandem mass spectrometry (LC-MS/MS) assay for simultaneous measurement of salivary testosterone and cortisol in healthy men for utilization in the diagnosis of late-onset hypogonadism in males, *Endocr. J.*, 2009, **56**, 1083–1093.
179. W. J. Shih, P. S. Bachorik, J. A. Haga, G. L. Myers and E. A. Stein, Estimating the long-term effects of storage at-70 degrees C on cholesterol, triglyceride, and HDL-cholesterol measurements in stored sera, *Clin. Chem.*, 2000, **46**, 351–364.

180. H. K. Kley, R. Schlaghecke and H. L. Kruskemper, Stability of steroids in plasma over a 10-year period, *J. Clin. Chem. Clin. Biochem.*, 1985, **23**, 875–878.
181. E. Matteucci, E. Biasci and O. Giampietro, Advanced oxidation protein products in plasma: stability during storage and correlation with other clinical characteristics, *Acta Diabetol.*, 2001, **38**, 187–189.
182. M. J. Evans, J. H. Livesey, M. J. Ellis and T. G. Yandle, Effect of anticoagulants and storage temperatures on stability of plasma and serum hormones, *Clin. Biochem.*, 2001, **34**, 107–112.
183. A. H. Garde, Å. M. Hansen and J. Kristiansen, Evaluation, including effects of storage and repeated freezing and thawing, of a method for measurement of urinary creatinine, *Scand. J. Clin. Lab. Invest.*, 2003, **63**, 521–524.
184. Å. M. Hansen, A. H. Garde and R. Persson, Sources of biological and methodological variation in salivary cortisol and their impact on measurement among healthy adults: a review, *Scand. J. Clin. Lab. Invest.*, 2008, **68**, 448–458.

CHAPTER 8
Micronuclei for Human Biomonitoring

MICHELINE KIRSCH-VOLDERS AND
ILSE DECORDIER*

Vrije Universiteit Brussel, Laboratorium voor Cellulaire Genetica,
Pleinlaan 2, 1050 Brussels, Belgium

8.1 Introduction

Micronuclei (MN) were originally identified and described as Howell–Jolly bodies in erythrocytes. In the following decades similar structures were described by a number of cytogeneticists following *in vitro* irradiation of cells.[1,2] In the early 1970s the term micronucleus test was suggested for the first time by Boller and Schmidt[3] and Heddle,[4] who showed that this assay provided a simple method to detect the genotoxic potential of mutagens after *in vivo* exposure of animals using bone marrow erythrocytes. A few years later Countryman and Heddle[5] showed that peripheral blood lymphocytes could be also used for the micronucleus approach and they recommended using micronuclei as a biomarker in testing schemes. Nowadays the scoring of micronuclei is widely used in biomonitoring of human populations exposed to different environmental, occupational or lifestyle factors and in genotoxicity testing because of its capacity to detect both clastogenic and aneugenic events, simplicity of scoring, accuracy, multi-potentiality, wide applicability in different cell types and amenability to automation. This chapter aims at providing a brief review of our present understanding of MN as a biomarker of early genetic effects, including their formation, fate and predictivity for cancer. In addition the different methodologies, automation of MN scoring and applications of the MN assay in human biomonitoring are highlighted.

Issues in Toxicology No. 10
Biomarkers and Human Biomonitoring, Volume 2: Selected Biomarkers of Current Interest
Edited by Lisbeth E. Knudsen and Domenico Franco Merlo

Published by the Royal Society of Chemistry, www.rsc.org

8.2 MN as a Biomarker for Early Genetic Effects

8.2.1 Origin and Mechanisms of MN Formation

The MN are small, extranuclear bodies that arise in dividing cells from acentric chromosome/chromatid fragments or whole chromosomes/chromatids that lag behind in anaphase and are not included in the daughter nuclei at telophase, but are encapsulated into a separate, smaller nucleus, a micronucleus (MN).[6,7] MN can be induced either spontaneously or by mutagens (Figure 8.1A). Examples of spontaneously formed MN are the chromosome loss found in normal neural progenitor cells,[8] or MN that occurs as a defence mechanism of the cell to remove extra chromosomal double minutes (dmin). These micronuclei result from gene amplification *via* breakage–fusion–bridge (BFB) cycles in which amplified DNA is localized selectively to specific sites at the periphery of the nucleus and eliminated via nuclear budding (NBUD) during the S-phase of the cell cycle.[6,7] MN originating from lagging acentric chromosomes or chromatid fragments are caused by misrepair of DNA breaks or unrepaired DNA breaks and are induced mainly by clastogens. Malsegregation of whole

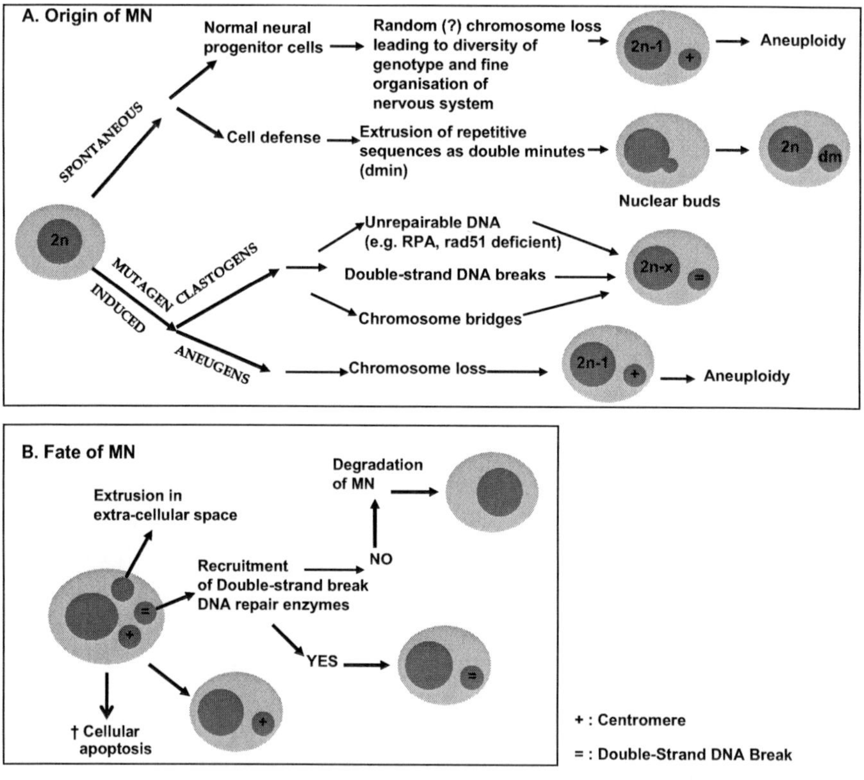

Figure 8.1 A. Model of the different routes leading to MN formation (see text for explanation). B. Some of the possible post-mitotic fates of MN (see text for explanation), adapted from ref. 11.

chromosomes at anaphase may lead to MN formation as a result of defects in kinetochore proteins or assembly, dysfunctional spindle and defective anaphase checkpoint genes, or hypomethylation of repeat sequences in centromeric and pericentromeric DNA, and are mainly induced by aneugens. Nucleoplasmic bridges (NPB) originate from dicentric chromosomes, which may occur as a result of misrepair of DNA breaks, telomere end fusions, and may also be observed when defective separation of sister chromatids at anaphase occurs owing to a failure of decatenation. NBUD represent the process of elimination of amplified DNA, DNA repair complexes and possibly excess chromosomes from aneuploid cells (Figure 8.2).[6,7]

8.2.2 Fate of MN

The post-mitotic fate of MN is poorly understood and different possibilities have been described:[6,9] i) retention within the cell's cytoplasm as an extra-nuclear entity, when MN may complete one or more rounds of DNA/chromosome replication; ii) re-incorporation into the main nucleus; iii) elimination of the micronucleated cells by apoptosis; iv) expulsion from the cell (Figure 8.1B). The ability of micronuclear chromatin to exert proper DNA replication, transcription and repair is dependent on the specific genetic content of the micronucleus as well as the functionality of its own envelope.[9,10] As far as apoptosis is concerned, one can consider either that the micronucleated cell as such is undergoing apoptosis or that the apoptotic process is restricted to the MN. Decordier *et al.*[11,12] demonstrated that micronucleated cells induced by aneugens constitute a strong apoptotic signal. In addition, they suggested a role of caspase-3, besides its function as an effector caspase in the apoptotic pathway, in MN formation. MN-harbouring cells tend to die during cell progression with a greater frequency than cells without MN, and malfunctioning of gene expression in the micronuclei might trigger apoptosis.[13] The induction of apoptosis in the MN itself is considered to be dependent on the presence or absence of DNA repair and/or DNA replication checkpoint machinery.

8.3 Cancer Predictivity in Adults

Depending on the fate of the MN, the ploidy of the daughter cells may vary,[6] therefore micronucleation leads potentially to a large variety of DNA/chromosome contents, with the known consequences of these imbalanced genotypes in terms of cell transformation and cancer risk. Therefore MN are indicators for fixed genomic damage in cells.

As far as baseline MN frequencies are concerned, the presence of an association between MN induction and cancer development in untreated cancer patients has been reported in a large number of studies.[14] A prospective study performed in the framework of the International Human Micronucleus (HUMN) Project (www.humn.org) demonstrated a significantly higher prospective cancer risk [relative risk (RR) 1.67, $p = 0.002$] in those diagnosed with MN frequencies in the medium or high tertile as compared to those in the low

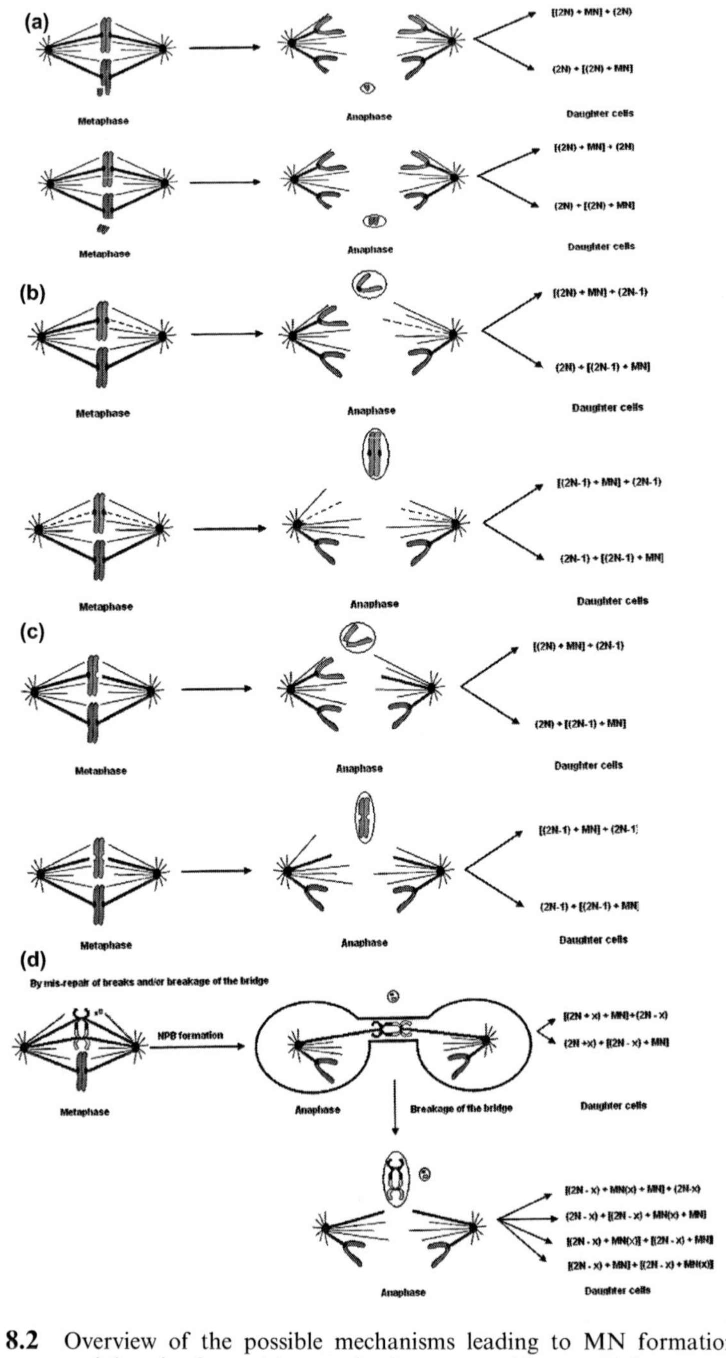

Figure 8.2 Overview of the possible mechanisms leading to MN formation. MN mainly arise from: (a) acentric chromosome/chromatid fragments resulting from DNA breakage events; (b) whole chromosomes/chromatids that lag behind in anaphase owing to misattachment of tubulin fibres on kinetochores or tubulin depolymerization; (c) defects in centromeric DNA, in kinetochore proteins or in kinetochore assembly; (d) MN can also arise as a result of NPB formation/breakage. Adapted from refs. 6 and 7.

MN frequency tertile.[15] The evidence that baseline frequencies of MN are predictive for cancer risk was confirmed by Murgia *et al.*[16] in a nested case–control study performed 14 years after the original recruitment and based on different cancer sites. As far as induced MN frequencies are concerned, convincing evidence that elevated levels of MN in peripheral lymphocytes are associated with cancer has been demonstrated in lung cancer[17,18] and in pancreatic cancer.[19]

Besides cancer, the MN frequency in lymphocytes measured using the CBMN assay has been associated prospectively with increased pregnancy complications[20] and cardiovascular disease mortality.[21,22]

8.4 Methodologies

8.4.1 In Human Lymphocytes

Micronuclei can be assessed or scored easily in erythrocytes, lymphocytes and exfoliated epithelial cells (*e.g.* oral, urothelial, nasal) to obtain a measure of genome damage induced *in vivo*. The standard *ex vivo*/*in vitro* MN test is usually performed in lymphocytes. Given that the formation of a MN requires a nuclear division, it is necessary to be able to distinguish dividing cells from resting cells. The cytokinesis-block micronucleus (CBMN) methodology developed by Fenech and Morley[23] uses cytochalasin B to identify cells that have divided in culture. Cytochalasin B is an inhibitor of actin polymerization which is required for the formation of the microfilament ring that constricts the cytoplasm between the daughter nuclei during cytokinesis. The CBMN methodology allows distinction between binucleated cells (BN; cells that have divided once in culture) and mononucleated cells (MONO; cells that did not divide or escaped the cytokinesis-block). A schematic representation of the classical *ex vivo*/*in vitro* CBMN test is given in Figure 8.3. It has been suggested that MN present in binucleated cells (MNBN) as well as in mononucleated cells (MNMONO) should be taken into account when performing the CBMN assay.[24] MNMONO indicate chromosome damage that was present *in vivo* before the start of culture and give an estimation of chromosome/genome mutations accumulated over many years, while MNBN contain pre-existing MN plus lesions that are expressed as MN during *in vitro* culture.

A major advantage of the CBMN assay lies in its ability to detect both clastogenic and aneugenic events. The use of FISH with probes labeling the pan (peri)centromeric region of chromosomes allows distinction between micronuclei that contain a whole chromosome (centromere positive micronucleus) and an acentric chromosome fragment (centromere negative micronucleus).[25] In addition, the CBMN assay can provide a measure of genotoxicity and cytotoxicity: nucleoplasmic bridges (NPB, a marker of chromosome rearrangement), nuclear buds (NBUD, a marker of gene amplification), cell division inhibition (by estimation of the nuclear division index), necrosis and apoptosis. For this reason, the CBMN test can be considered as a "cytome" assay

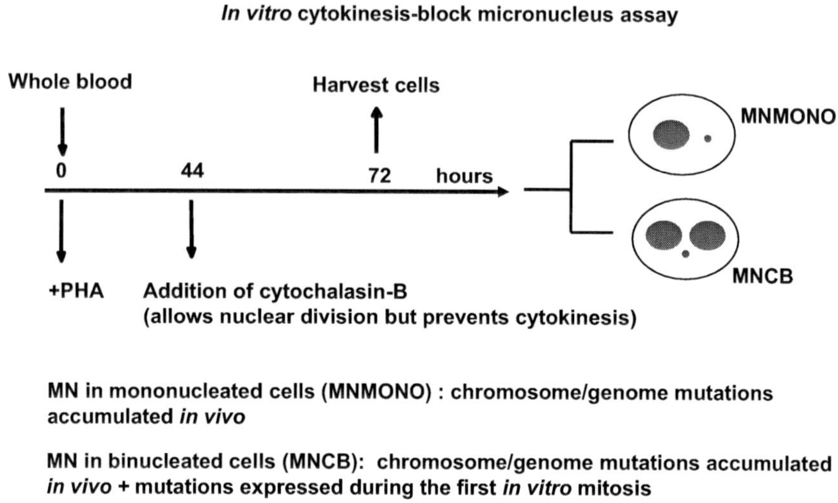

Figure 8.3 *In vitro* cytochalasin-B micronucleus assay: methodology. In a classical test, human lymphocytes are cultured in the presence of phytohaemagglutinin to stimulate mitosis. After 44 hours, cytochalasin-B is added to the culture. The use of this inhibitor of actin polymerization will block cytokinesis, allowing distinction between binucleated cells (cells that have divided once in culture) and mononucleated cells (cells that did not divide or escaped the cytokinesis-block). At 72 hours the cells are harvested onto microscope slides, fixed and stained. Adapted from ref. 6.

covering chromosome instability, mitotic dysfunction, cell proliferation and cell death.[26] Major steps in the validation of the CBMN assay for human biomonitoring were achieved by the HUMN project, which examined the major confounding factors (culture conditions, scoring criteria, age, smoking, genotype, exposure) that influence MN induction. The CBMN protocol and scoring criteria are now standardized; the effects of age, gender and smoking status have been defined.[27]

8.4.2 In Buccal Cells

In addition to the MN assay in lymphocytes, the MN assay in exfoliated buccal cells provides a complementary method for measuring MN in a tissue that is easily accessible and does not require tissue culture. Buccal cells are the first barrier for the inhalation or ingestion route and are capable of metabolizing proximate carcinogens to reactive products.[28] The MN assay in exfoliated buccal cells has been used since the 1980s to demonstrate cytogenetic effects of environmental and occupational exposures, lifestyle patterns and diseases. However, important issues remain about the characteristics of MN and other nuclear abnormalities, the basic biology explaining the appearance of various cell types in samples of buccal mucosa, and the effects of diverse staining procedures and scoring criteria in different laboratories and their relationship to disease states and outcomes.[28] Therefore, the HUMN coordinating group started a new international validation project for the buccal cell MN assay that is similar to the project previously performed using human lymphocytes: $HUMN_{XL}$, *i.e.* human MN assay in exfoliated buccal cells. The planned research will explore sources of variability in the assay (*e.g.* between laboratories and scorers, as well as inter- and intra-individual differences among subjects) and resolve key technical issues, such as the method of buccal cell collection and staining, and optimal criteria for classification of normal and degenerated cells and for scoring MN and other abnormalities. Recently, the protocol for the buccal micronucleus cytome (BMCyt) assay used for studying DNA damage and chromosomal instability has been described in detail (Figure 8.4). It includes buccal cell collection, generation of a single-cell suspension, slide preparation using cytocentrifugation, fixation and staining, and the scoring criteria for micronuclei and other nuclear anomalies.[29]

8.5 Automation of MN Scoring for Biomonitoring

Automation of MN analysis allows a quicker and more reliable detection, minimizing subjective judgement and individual scoring skills, and allowing

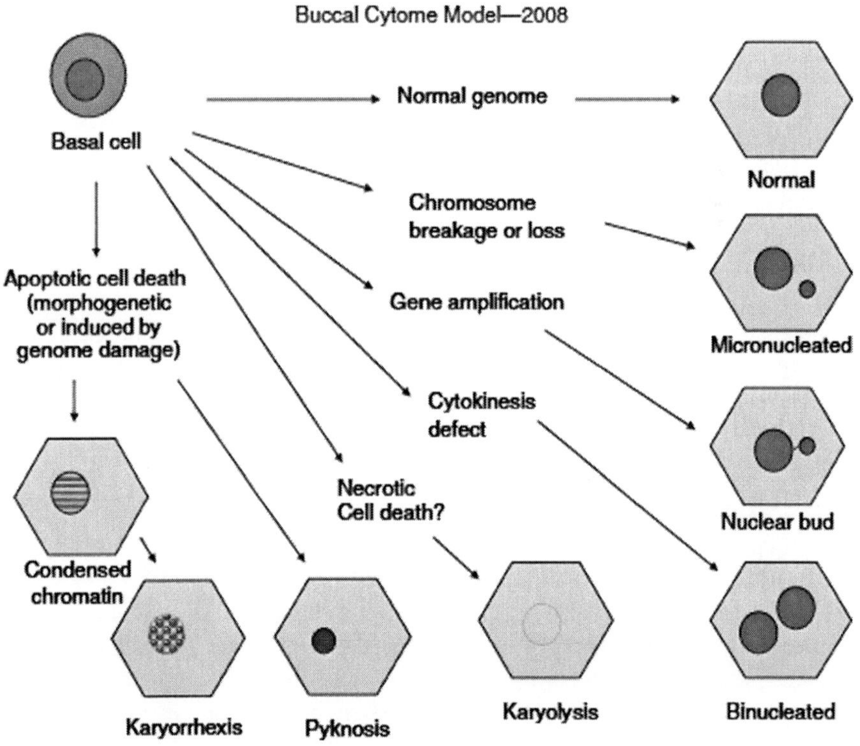

Figure 8.4 Diagrammatic representation of various cell types scored in the BMCyt assay, adapted from ref. 29.

applicability on a large scale for human biomonitoring. An automated image analysis system for the CBMN assay providing robust, reliable and reproducible results should fulfil a set of essential prerequisites including a well defined detection algorithm, applicable to human lymphocytes, that discriminates among mono-, bi- and polynucleated cells, MN scoring according to HUMN scoring criteria, it should be thoroughly validated, the false positive and false negative rates should be as low as possible, and determination of cell proliferation should be available.[30] Currently, there are two systems for routine automated analysis of MN on the market: Metafer MNScore (by MetaSystems GmbH Altlussheim, Germany)[31–33] and PathFinder™Cellscan™ (by IMSTAR, Paris, France). The latter was developed and validated by the authors in collaboration with IMSTAR,[34] in the framework of the EU Integrated Project NewGeneris (www.newgeneris.org), and fulfils the above criteria. The main difference between the two systems lies in the detection algorithms. The algorithms applied for the Metafer MNScore only detect binucleated entities consisting of two similar nuclei close to each other but completely separated, with or without MN.[31,32] The specific algorithms described for the IMSTAR system start from the cell as a detection unit: first, the cells and nuclei are detected, then the MN are searched for in the detected cells.[34] This allows identifications of mono-, bi- and polynucleated cells, and MN in these different sub-populations of cells, and hence the assessment of cell proliferation through the nuclearity index, which is important for an efficient assessment of the genetic damage in human biomonitoring because these criteria are indicative of immune function and cytotoxicity.[26] Scoring of mononucleated cells also allows the detection of micronuclei in these cells (MNMONO).[24] Although the automated image analysis system of MetaSystems has already been applied in research related to mutagen sensitivity phenotypes in cancer risk, radiation biodosimetry and biomonitoring studies of air pollution, large differences in lymphocyte culture conditions and processing of samples are observed.[33] When developing the IMSTAR automated MN image analysis system, we established a well-standardized slide preparation protocol,[34] which ensures high reproducibility and can be used to perform an unambiguous and consistent image analysis.

8.6 Applications of the MN Assay for Human Biomonitoring

8.6.1 Occupational and Environmental Exposure

The use of MN as a measure of early genotoxic effects can greatly improve risk assessment and surveillance of populations exposed to mutagens/carcinogens. MN are biomarkers that have been widely used for studying the most important occupational and environmental hazards to public health that have occurred in the past few decades.

The exposures studied most frequently include pesticides, radiation exposure of occupational, medical and accidentally exposed individuals, styrene, polycyclic aromatic hydrocarbons (PAHs), benzene, ethylene oxide, formaldehyde, welding fumes, arsenic, metals, environmental pollution, and cytostatics.[35–38] Most studies evaluated MN in peripheral blood lymphocytes (PBL), although an increasing number of studies are performed on exfoliated epithelial cells. Application of MN as cytogenetic endpoints in human biomonitoring has been extensively performed through single studies,[39] which are usually small in size. To increase the statistical power of such single study approaches several pooled/meta-analyses have been undertaken over recent years, allowing a better assessment of the questions raised by the preliminary single biomonitoring studies (*e.g.* the association with cancer risk, and the impact of potential confounders on the induction of cellular phenotypes).[39] Application of the MN assay in population biomonitoring studies requires a deep understanding of how age, gender, diet and lifestyle factors (*e.g.* exercise, alcohol, smoking and recreational drugs) may influence the MN frequency in PBL. It is clear that these factors can influence MN frequency significantly. Proper control for these factors is required to enable better measurement of the impact of other conditions, such as environmental exposure to genotoxins or a susceptible genetic background, on MN frequency.[40]

8.6.2 Micronuclei in Newborns and Children

The MN assay is increasingly used in children because of its public health significance: responses of children to environmental toxicants and carcinogens can differ from those observed in adults, both in the severity and in the nature of the effect, owing to their unique exposures and special vulnerability to environmental toxicants and carcinogens.[41] The MN assay has been applied for monitoring environmental exposures such as pollution, radiation, industrial toxicants and various medical treatments of children and to assess genotoxicity associated with diseases such as leukaemia.[42] The frequencies of MN in newborns are quite low when compared to those in adults. Given that MN formation is dependent on cell division, and that MN can accumulate in a given cell up to the death of this cell, it is expected that frequencies of micronuclei in T-lymphocytes in non-exposed newborns and infants (until the age of approximately 5 years) will increase to a plateau level defined by the turn-over time of the T-lymphocytes. As far as newborns are concerned, MN in lymphocytes of umbilical blood reflect the average fetal genome instability plus *in utero* exposure to clastogens/aneugens. Therefore, four major facts need to be considered: (i) the T-lymphocytes to be analysed by the MN assay are circulating in peripheral blood for only 6 months; (ii) the response of T-lymphocytes to phytohaemagglutinin stimulation in umbilical cord blood is less efficient than in peripheral blood from adults; (iii) genome stability may be different during fetal life; and (iv) the baseline MN frequency is relatively low in newborns and its assessment requires large cohorts and cell sample counts.[42] For MN studies in children, lymphocytes are most commonly used but recently more investigators have become interested in using exfoliated cells that can be collected non-invasively. Most studies are consistent in concluding that environmental pollutants and radiation exposures lead to an increase in the MN frequency in children; effects of medical treatments are less clear. Various factors of importance to MN variability in children include tissue specificity, age, sex, health status and environmental and dietary factors. Despite the recent progress in MN assay in children, more studies are needed to explain sex, age and genotype differences in baseline MN levels and the changes in response to genotoxicants. Moreover, owing to physiological differences and the age of circulating T-lymphocytes, it is not yet assessed whether MN frequencies can be interpreted in the same way as for adults and whether they are predictive for cancer.

8.7 Conclusion and Future Perspectives

The *in vitro* MN test detects micronuclei (MN) as small, extranuclear bodies that arise in dividing cells from acentric chromosome/chromatid fragments or whole chromosomes/chromatids that lag behind in anaphase and are not included in the daughter nuclei at telophase. The different mechanisms leading to the formation of MN are well understood; their possible post-mitotic fate is less evident. MN have become an important cytogenetic biomarker in the biomonitoring of human populations for *in vivo* genotoxic exposure. The MN assay has the capacity to detect both aneugens and clastogens, shows simplicity of scoring, is widely applicable in different cell types, is internationally validated, is predictive for cancer, and its statistical power is increased by the recent high throughput methodologies. Despite all these achievements, there are still several challenges to be met: an important issue is the comprehensive micronucleus cytome approach, which is increasingly being adopted because it enables all major nuclear anomalies and cytotoxicity events to be captured simultaneously. However, validation for other biomarkers, in addition to MN, in the cytome system and their automated scoring is still needed. Furthermore larger and/or longer studies are required to verify the results of previous studies concerning the association of MN frequency with cancer and other diseases. In addition, future studies should explore the relationships between MN expression and transcriptome, metabolome and proteome profiles to unravel the underlying molecular mechanisms that correlate with this biomarker of DNA damage. These "omics" data could provide valuable information on the likely origin of MN when the exposure profile is unknown or difficult to ascertain.

Acknowledgements

This work was supported by the EU research programmes "The Detection and Hazard Evaluation of Aneugenic Chemicals" (ENV4-CT97-0471), "Protection of the European Population from Aneugenic Chemicals (PEPFAC)" (QLK4-CT-2000-00058), Cancer Risk Biomarkers (Contract no. QLK4-2000-00628), ECNIS (Environmental Cancer Risk, Nutrition and Individual Susceptibility), a network of excellence operating within the European Union 6th Framework Program, Priority 5: "Food Quality and Safety" (Contract No 513943) and by the EU Integrated Project NewGeneris ("Newborns and Genotoxic exposure risk"), 6th Framework Programme, Priority 5: Food Quality and Safety (Contract no. FOOD-CT-2005-016320). NewGeneris is the acronym of the project, the Belgian Science Policy [contract number SD/HE/02B]; OZR-VUB.

References

1. J. M. Thoday, The effect of ionizing radiation on the broad bean root. Part IX. Chromosome breakage and the lethality of ionizing radiations to the root meristem, *Br. J. Radiol.*, 1951, **24**, 276–572.
2. H. J. Evans, G. J. Neary and F. S. Williamson, The relative biological efficiency of single doses of fast neutrons and gamma rays in *Vicia faba* roots and the effect of oxygen, Part II. Chromosome damage; the production of micronuclei, *Int. J. Rad. Biol.*, 1959, **1**, 230–240.
3. K. Boller and W. Schmid, Chemical mutagenesis in mammals. The Chinese hamster bone marrow as an *in vivo* test system. Hematological findings after treatment with trenimon, *Humangenetik*, 1970, **1**, 35–54.
4. J. A. Heddle, A rapid *in vivo* test for chromosome damage, *Mutat. Res.*, 1973, **18**, 187–190.
5. P. I. Countryman and J. A. Heddle, The production of micronuclei from chromosome aberrations in irradiated cultures of human lymphocytes, *Mutat. Res.*, 1976, **4**, 321–332.
6. R. Mateuca, N. Lombaert, P.V. Aka, I. Decordier and M. Kirsch-Volders, Chromosomal changes: induction, detection methods and applicability in human biomonitoring, *Biochimie*, 2006, **88**, 1515–1531.
7. M. Fenech, M. Kirsch-Volders, A. T. Natarajan, J. Surrales, J. Croft, J. Parry and H. Norppa, Molecular mechanisms of micronucleus, nucleoplasmic bridge and nuclear bud formation in human cells, *Mutagenesis*, 2011, **26**, 125–132.
8. A. H. Yang, D. Kaushal, S. K. Rehen, K. Kriedt, M. A. Kingsbury, M. J. McConnell and J. Chun, Chromosome segregation defects contribute to aneuploidy in normal neural progenitor cells, *J. Neurosci.*, 2003, **23**, 10454–10562.
9. M. Terradas, M. Martín, L. Tusell and A. Genescà, Genetic activities in micronuclei: is the DNA entrapped in micronuclei lost for the cell? *Mutat. Res.*, 2010, **705**, 60–67.
10. M. Kirsch-Volders, G. Plas, A. Elhajouji, M. L. Gonzalez, K. Vande Loock and I. Decordier, The *in vitro* MN assay in 2010: origin and fate, biological significance, protocols, high throughput methodologies and toxicological relevance, *Arch. Toxicol.*, 2011, in press.
11. I. Decordier, L. Dillen, E. Cundari and M. Kirsch-Volders, Elimination of micronucleated cells by apoptosis after treatment with inhibitors of microtubules, *Mutagenesis*, 2002, **17**, 337–344.

12. I. Decordier, E. Cundari, L. and M. Kirsch-Volders, Influence of caspase activity on micronuclei detection: a possible role for caspase-3 in micronucleation, *Mutagenesis*, 2005, **20**, 173–179.
13. K. Utani, Y. Kohno, A. Okamoto and N Shimizu, Emergence of micronuclei and their effects on the fate of cells under replication stress, *PLoS One*, 2010, **5**, e10089.
14. S. Bonassi, R. El-Zein, C. Bolognesi and M. Fenech, Micronuclei frequency in peripheral blood lymphocytes and cancer risk: evidence from human studies, *Mutagenesis*, 2011, **26**, 93–100.
15. S. Bonassi, A. Znaor, M. Ceppi, C. Lando, W. P. Chang, N. Holland, M. Kirsch-Volders, E. Zeiger, S. Ban, R. Barale, M. P. Bigatti, C. Bolognesi, A. Cebulska-Wasilewska, E. Fabianova, A. Fucic, L. Hagmar, G. Joksic, A. Martelli, L. Migliore, E. Mirkova, M. R. Scarfi, A. Zijno, H. Norppa and M. Fenech, An increased micronucleus frequency in peripheral blood lymphocytes predicts the risk of cancer in humans, *Carcinogenesis*, 2007, **28**, 625–631.
16. E. Murgia, M. Ballardin, S. Bonassi, A. M. Rossi and R. Barale, Validation of micronuclei frequency in peripheral blood lymphocytes as early cancer risk biomarker in a nested case-control study, *Mutat. Res.*, 2008, **39**, 27–34.
17. R. A. El-Zein, M. B. Schabath, C. J. Etzel, M. S. Lopez, J. D. Franklin and M. R. Spitz, Cytokinesis-blocked micronucleus assay as a novel biomarker for lung cancer risk, *Cancer Res.*, 2006, **66**, 6449–6464.
18. R. A. El-Zein, M. Fenech, M. S. Lopez, M. R. Spitz and C. J. Etzel, Cytokinesis-blocked micronucleus cytome assay biomarkers identify lung cancer cases amongst smokers, *Cancer Epidemiol. Biomarkers Prev.*, 2008, **17**, 111–119.
19. P. Chang, Y. Li and D. Li, Micronuclei levels in peripheral blood lymphocytes as a potential biomarker for pancreatic cancer risk, *Carcinogenesis*, 2011, **3**, 210–215.
20. D. L. Furness, G. A. Dekker, W. M. Hague, T. Y. Khong and M. F. Fenech, Increased lymphocyte micronucleus frequency in early pregnancy is associated prospectively with pre-eclampsia and/or intrauterine growth restriction, *Mutagenesis*, 2010, **25**, 489–398.
21. E. Murgia, V. Maggini, R. Barale and A. M. Rossi, Micronuclei, genetic polymorphisms and cardiovascular disease mortality in a nested case-control study in Italy, *Mutat. Res.*, 2007, **621**, 113–118.
22. C. Federici, N. Botto, S. Manfredi, A. Rizza, M. Del Fiandra and M. G. Andreassi, Relation of increased chromosomal damage to future adverse cardiac events in patients with know coronary artery disease, *Am. J. Cardiol.*, 2008, **102**, 1296–1300.
23. M. Fenech and A. A. Morley, Measurement of micronuclei in lymphocytes, *Mutat. Res.*, 1985, **147**, 29–36.
24. M. Kirsch-Volders and M. Fenech, Inclusion of micronuclei in non-divided mononuclear lymphocytes and necrosis/apoptosis may provide a more

comprehensive cytokinesis block micronucleus assay for biomonitoring purposes, *Mutagenesis*, 2001, **16**, 51–58.
25. I. Decordier, R. Mateuca and M. Kirsch-Volders, Micronucleus assay and labeling of centromeres with FISH technique, *Methods Mol. Biol.*, 2011, **691**, 115–136.
26. M. Fenech, Cytokinesis-block micronucleus cytome assay, *Nat. Protoc.*, 2007, **2**, 1084–1104.
27. M. Fenech, N. Holland, E. Zeiger, W. P. Chang, S. Burgaz, P. Thomas, C. Bolognesi, S. Knasmueller, M. Kirsch-Volders and S. Bonassi, The HUMN and HUMNxL international collaboration projects on human micronucleus assays in lymphocytes and buccal cells – past, present and future, *Mutagenesis*, 2011, **26**, 239–245.
28. N. Holland, C. Bolognesi, M. Kirsch-Volders, S. Bonassi, E Zeiger, S. Knasmueller and M. Fenech, The micronucleus assay in human buccal cells as a tool for biomonitoring DNA damage: the HUMN project perspective on current status and knowledge gaps, *Mutat. Res.*, 2008, **659**, 93–108.
29. P. Thomas and M. Fenech, Buccal micronucleus cytome assay, *Nat. Protoc.*, 2009, **4**, 825–837.
30. I. Decordier, A. Papine, K. Vande Loock, G. Plas, F. Soussaline and M. Kirsch-Volders, Automated image analysis of MN by IMSTAR for biomonitoring, *Mutagenesis*, 2011, **26**, 163–168.
31. C. Schunck, T. Johannes, D. Varga and A. Plesch, New developments in automated cytogenetic imaging: unattended scoring of dicentric chromosomes, micronuclei, single cell gel electrophoresis, and fluorescence signals, *Cytogenet. Genome Res.*, 2004, **104**, 383–389.
32. D. Varga, T. Johannes, S. Jainta, S. Schuster, U. Schwarz-Boeger, M. Kiechle, B. Patino Garcia and W. Vogel, An automated scoring procedure for the micronucleus test by image analysis, *Mutagenesis*, 2004, **19**, 391–397.
33. A. Rossnerova, M. Spatova, C. Schunck and R. J. Sram, Automated scoring of lymphocyte micronuclei by the MetaSystems Metafer image cytometry system and its application in studies of human mutagen sensitivity and biodosimetry of genotoxin exposure, *Mutagenesis*, 2011, **26**, 169–175.
34. I. Decordier, A. Papine, G. Plas, S. Roesems, K. Vande Loock, J. Moreno-Palomo, E. Cemeli, D. Anderson, A. Fucic, R. Marcos, F. Soussaline and M. Kirsch-Volders, Automated image analysis of cytokinesis-blocked micronuclei: an adapted protocol and a validated scoring procedure for biomonitoring, *Mutagenesis*, 2009, **24**, 85–93.
35. S. Bonassi, D. Ugolini, M. Kirsch-Volders, U. Strömberg, R. Vermeulen and J. D. Tucker, Human population studies with cytogenetic biomarkers: review of the literature and future prospectives, *Environ. Mol. Mutagen.*, 2005, **45**, 258–270.
36. A. Vral, M. Fenech and H. Thierens, The micronucleus assay as a biological dosimeter of *in vivo* ionising radiation exposure, *Mutagenesis*, 2011, **26**, 11–17.

37. R. A. Mateuca, M. Roelants, G. Iarmarcovai, P. V. Aka, L. Godderis, A. Tremp, S. Bonassi, M. Fenech, J. L. Bergé-Lefranc and M. Kirsch-Volders, hOGG1(326), XRCC1(399) and XRCC3(241) polymorphisms influence micronucleus frequencies in human lymphocytes *in vivo*, *Mutagenesis*, 2008, **23**, 35–41.
38. C. Bolognesi, A. Creus, P. Ostrosky-Wegman and R. Marcos, Micronuclei and pesticide exposure, *Mutagenesis*, 2011, **26**, 19–26.
39. R. A. Mateuca, I. Decordier and M. Kirsch-Volders, Cytogenetic methods in human biomonitoring: principles and uses, *Methods Mol. Biol.*, 2011, in press.
40. M. Fenech and S. Bonassi, The effect of age, gender, diet and lifestyle on DNA damage measured using micronucleus frequency in human peripheral blood lymphocytes, *Mutagenesis*, 2011, **26**, 43–49.
41. P. J. Landrigan, C. A. Kimmel, A. Correa and B. Eskenazi, Children's health and the environment: public health issues and challenges for risk assessment, *Environ. Health Perspect.*, 2004, **112**, 257–265.
42. N. Holland, A. Fucic, D. F. Merlo, R. Sram and M. Kirsch-Volders, Micronuclei in neonates and children: effects of environmental, genetic, demographic and disease variables, *Mutagenesis*, 2011, **26**, 51–56.

CHAPTER 9

Biomarkers of Individual Susceptibility: Genetic Polymorphisms and their Interplay with Micronucleus Frequencies

ILSE DECORDIER,* KIM VANDE LOOCK AND MICHELINE KIRSCH-VOLDERS

Vrije Universiteit Brussel, Laboratorium voor Cellulaire Genetica, Pleinlaan 2, 1050 Brussels, Belgium

9.1 Introduction

Understanding the genetic bases of individual susceptibility to environmental/occupational genotoxins can lead to a more precise prediction of human health problems and to more effective prevention of disease.[1] Therefore, intensive research has been focused on the study of genetic polymorphisms as an important component of the individual susceptibility phenomenon. At the most basic level, genetic polymorphisms represent differences in the base pair sequence of deoxyribonucleic acid (DNA). These differences in the base pair sequence of DNA can result from a variety of physical events and include transitions, transversions, missense mutations, or nonsense mutations.[2] Investigation of the human genome has revealed that the most common type of sequence variation consists of single nucleotide polymorphisms (SNPs).[3,4] SNPs are genomic loci at which two alternative bases are present with appreciable frequency (greater than 1%).[1] More than 10 million SNPs have been

Issues in Toxicology No. 10
Biomarkers and Human Biomonitoring, Volume 2: Selected Biomarkers of Current Interest
Edited by Lisbeth E. Knudsen and Domenico Franco Merlo

Published by the Royal Society of Chemistry, www.rsc.org

identified across the genome. Most (99%) of these polymorphisms are biologically silent, while the effects and biological significance of the remainder vary according to their position within the genome.[5] Moreover, it is well know that genotypic polymorphism of DNA repair genes is associated with the risk of cancer, but the functional role of genetic variants contributing to diseases such as cancer is poorly known.[6] An important step in understanding the role that SNPs play in disease is to consider the complexity of the relationships involved and their impact on the genome. Micronuclei (MN) have been widely used to assess genotype–phenotype associations. The advantages of the MN assay as a valuable tool for this approach lie in its wide application in human biomonitoring studies, the wide range of events detected by this assay,[7] its extensive validation and reliability for human biomonitoring, its cancer risk predictivity and the recent implementation of automated methods for MN detection.[8] The formation of MN in dividing cells is the result of chromosome breakage associated with unrepaired or misrepaired DNA lesions, or chromosome malsegregation due to mitotic malfunction, and it can be induced by exposure to clastogens or aneugens.[9] Recently, association studies have been published that link genotypes, which take into account inter-individual differences in response to genotoxic exposure, to the occurrence of MN, which quantify the extent of genetic damage. In this chapter we aimed to review briefly the link between the frequencies of MN arising from clastogenic and aneugenic effects and polymorphisms of genes implicated DNA repair, chromosome segregation, xenobiotic metabolism (activation/detoxification), and folate metabolism.

9.2 MN Derived from Clastogenic Effects and Genetic Polymorphisms

MN that harbour acentric chromosome/chromatid fragments are induced mainly by clastogens and result from double-strand DNA breaks (DSBs) induced either directly or indirectly by the conversion of single-strand DNA breaks (SSBs) into DSBs after cell replication. Therefore, genetic polymorphisms leading to misrepair of SSBs and DSBs may influence MN frequencies.[9] Base excision repair (BER) and double strand break repair (DSBR) are the main DNA repair pathways involved in the repair of SSBs and DSBs, respectively (Figure 9.1). Two examples of polymorphic genes that have been extensively studied in association with MN frequencies are *XRCC1* and *XRCC3*, which are involved in the repair of oxidized bases, SSBs and DSBs, respectively.[10] Moreover, despite some controversial results, genetic variants in these genes have been associated with cancer risk.[10]

XRCC1 plays an essential role in the repair of SSB and damaged bases through BER, via its ability to act as a scaffold upon which other proteins of the repair pathway can bind.[11] Three common polymorphisms in *XRCC1*, $Arg^{194}Trp$ ($XRCC1^{194}$), $Arg^{280}His$ ($XRCC1^{280}$) and $Arg^{399}Gln$ ($XRCC1^{399}$), have the potential to alter the ability of *XRCC1* to interact with some of its BER partners.[12] This may result in a decreased protein affinity or decreased

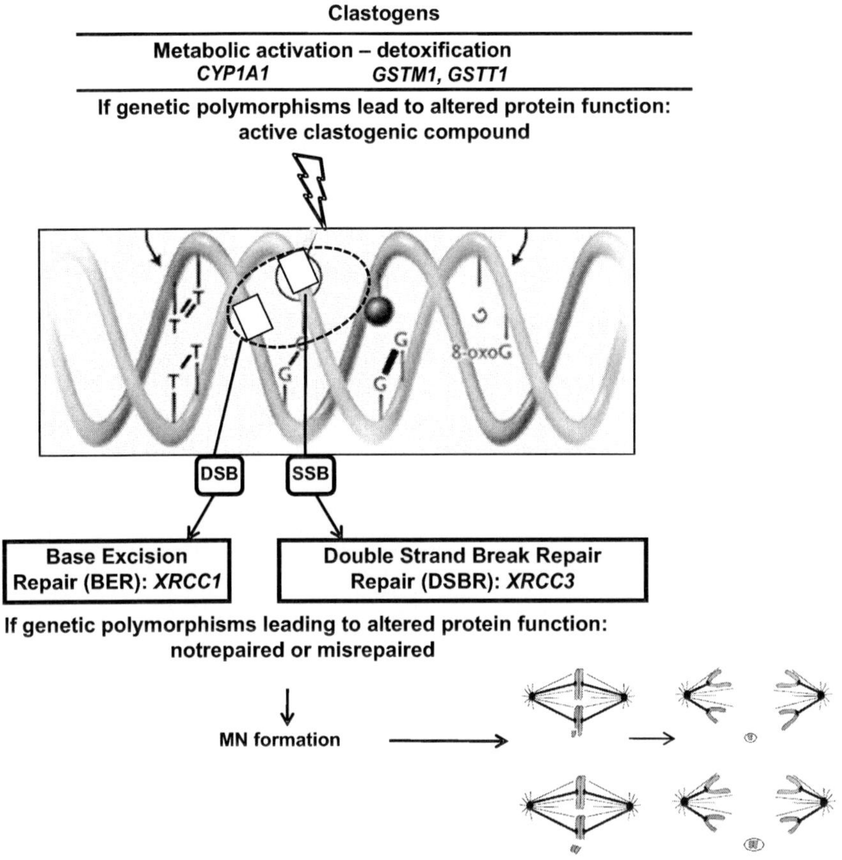

Figure 9.1 Model of how genetic polymorphisms in genes involved in xenobiotic metabolism and DNA repair can lead to MN containing chromatid/chromosome fragments after exposure to clastogens. The genes mentioned are examples discussed in the text.

binding to damaged DNA and ineffective DNA repair and can, therefore, influence genotoxic responses to clastogenic compounds, leading to the formation of MN (Figure 9.1). Several biomonitoring studies focusing on occupation/environmental exposure have demonstrated a significant influence of *XRCC1* polymorphisms on MNBN (MN in binucleated cells) induction and demonstrated that subjects occupationally or environmentally exposed and with *XRCC1* variant genotypes showed higher MN frequencies.[13–16] Mateuca *et al.*[10] observed, in a pooled analysis, elevated MN levels in smokers co-exposed to occupational genotoxins and carrying the heterozygous *XRCC1*[399] genotype, when compared to their non-smoking counterparts. Dhillon *et al.*[17] reviewed the evidence from human population studies linking genotype and MN frequency in PBLs, taking into consideration interactions with nutrient status and/or exposure to genotoxins. Many of the included studies reported an interaction between MN frequencies and certain genetic polymorphisms in DNA repair genes. However, this evidence was only strong enough for polymorphisms in *XRCC1* ($Arg^{280}His$), which appeared to be consistently associated with the MN formation. A study by Cheng *et al.*[18] evaluating DNA repair capacity and chromosomal damage measured as MN in coke-oven workers demonstrated that an association between DNA repair capacity and MN frequencies varied with respect to the genotype of $XRCC1^{399}$ polymorphism. This suggests that the variant genotype may be related to an impaired DNA repair capacity and thus may lead to increased MN frequencies that arise by chromosomal damage.

XRCC3 encodes a member of the RAD51-like proteins that participates in the homologous recombination (HR) repair of DSBs,[19] induced either directly or indirectly following replication of closely spaced SSBs.[20] The most thoroughly investigated polymorphism in *XRCC3* is $Thr^{241}Met$ ($XRCC^{241}$). Although the actual functional effect of this polymorphism is unknown, it has been reported that carriers of the variant *XRCC3* allele have high levels of DNA adducts in lymphocyte DNA, indicating that the polymorphism is associated with low DNA repair capacity.[21] Moreover, the variant *XRCC3* allele has been reported to be functionally deficient in the repair of chromosome aberrations induced by X-rays but not ultraviolet (UV) light, indicating that the variant genotype is defective in homologous recombination but not in nucleotide excision repair.[22] As for *XRCC1*, the influence of *XRCC3* polymorphism on MN frequencies has been evaluated and demonstrated in several studies; the results show that individuals carrying an *XRCC3 Met241* allele had higher MN levels when compared with homozygous *XRCC3 Thr241*.[23–25] It was also demonstrated that newborns carrying the variant $XRCC3^{241}$ genotype might be at higher risk for the induction of MN, when exposed to oxidative stress.[26]

Elevated frequencies of mononucleated cells bearing MN (MNMONO) have been observed in carriers of $XRCC3^{241}$ variant genotypes in occupationally exposed populations.[13,27] Given that MNMONO provide an estimate of the genomic instability accumulated *in vivo* before the start of the cell culture,[28] modulation of MNMONO levels by genetic polymorphisms reflects an *in vivo* exposure–genotype interaction. Unfortunately, most biomonitoring studies that address genotype–MN correlations restrict the MN scoring to binucleated cells (MNBN), and therefore data concerning the interplay among exposure, genotypes and MNMMONO induction is generally missing.

A major advantage of the cytokinesis-block micronucleus (CBMN) assay lies in its ability to distinguish between clastogenic and aneugenic effects when combined with fluorescence *in situ* hybridization using a pancentromeric DNA probe. This methodology discriminates between negatively labelled MN containing acentric chromosome fragments (centromere-negative MN, C–MN) and positively labelled MN (centromere-positive MN, C+MN) containing whole chromosomes.[29] Combination of the approach with genotype analysis in DNA repair genes would improve the sensitivity of human biomonitoring in individuals occupationally or environmentally exposed. Only few studies have applied this combined approach. Cho *et al.*[30] demonstrated elevated C–MN frequencies in industrial radiographers carrying the variant *XRCC3*^{241} genotype. Iarmarcovai *et al.*[31] evaluated the involvement of *XRCC1*^{199} and *XRCC3*^{241} polymorphisms on the centromere content of MN in welders, but did not find a significant different in C+MN or C–MN between the different genotypes.

For many studied polymorphisms,[17] often only a slight increase or marginally significant association of MN with SNPs is found in relation to occupational exposure and/or disease, owing to insufficient power to draw reliable conclusions. If the frequency of a variant allele in the general population is too low, there is always a greater need to increase the size of the study groups. An appropriate tool to summarize information from individual studies is pooled analysis, which increases the statistical power. Mateuca *et al.*[10] performed such a pooled analysis of five biomonitoring studies to investigated the influence of *hOGG1*^{326}, *XRCC1*^{399} and *XRCC3*^{241} gene polymorphisms on MN frequencies in human PBL, as measured by the CBMN assay. The findings indicated that single DNA repair gene polymorphisms are not likely to have a major impact on MN frequencies, but that the complex interplay between *hOGG1*^{326}, *XRCC1*^{399} and *XRCC3*^{241} genotypes and environmental factors modulates MN levels.[10]

9.3 MN derived from Aneugenic Effects and Genetic Polymorphisms

The association between MN derived from acentric chromosome/chromatid fragments and genetic polymorphisms in the DNA repair pathway has been extensively studied, and individual susceptibility to DNA damaging agents that cause structural chromosome aberrations has been linked to polymorphic variants of DNA repair genes. Similarly, the presence of MN harbouring whole chromosomes is likely to be influenced by genetic polymorphisms that control the maintenance of the correct centrosome number, the reactivity of aneugens, topoisomerases and cyclins, or the activity of cell cycle check points. However, very little information is available on the link between MN arising from chromosome loss, leading to aneuploidy, and genetic polymorphisms of genes involved in processes that ensure correct chromosome segregation, such as mitotic checkpoint, proper mitotic spindle function, centrosome function, all mechanisms that can give rise to MN.[9] Aneuploidy is the most common characteristic of human solid tumours and occurs early during tumorigenesis. Given that aberrant mitosis is a critical step in cancer development, and given the ubiquity of aneuploidy in human cancers,[32–34] it is reasonable to propose a hypothesis that links genes participating in chromosomal segregation with tumorigenesis. However, there is very little evidence linking common cancers and polymorphic genes involved in accurate chromosome segregation. Moreover, the association between these polymorphisms and the induction of MN has not yet been evaluated. Only recently have scientists investigated the association between genetic polymorphisms and cancer susceptibilty. Lo *et al.*[35] identified two SNPs in *Aurora-A* that were significantly associated with breast cancer risk. *Aurora-A* is a mitotic kinase-encoding gene of the serine/threonine kinase family. It is involved in the regulation of centrosomal duplication, which is considered to be a critical regulatory component of chromosomal segregation. Later, the same group found other SNPs in genes involved in the mitotic checkpoint to be associated with breast cancer risk: in *TTK* and *PTTG1*.[36] Olson *et al.*[37] performed the first epidemiological study that addressed the association between SNPs in genes related to the structure and function of chromosomes and breast cancer risk. Several genes were found to be associated with the risk of breast cancer. Additional functional analysis of these genes is necessary to explore the mechanisms by which they affect cancer risk.

As far as the genetic polymorphisms and MN arising from chromosome loss are concerned, some studies have addressed the possibility of an additional contribution of $XRCC3^{241}$ variants to the induction of MN arising from chromosome loss via malsegregation events. The variant $XRCC3^{241}$ protein has been associated with an increase in endoreduplication,[38] an increase in centrosome number,[39] and with higher levels of binucleated/multinucleated cells,[39] suggesting a possible involvement of this variant in chromosome loss, detectable as MN, and centrosome aberrations. Moreover, individual susceptibility to DNA damaging agents that induce structural chromosome aberrations, and thus MN resulting from clastogenic events, have been linked to polymorphic variants of DNA repair genes. Similar susceptibility variants have been reported only rarely for environmental/occupational agents that induce aneuploidy, such as spindle inhibitors.[40] Figure 9.2 presents a hypothetical model of how genetic polymorphisms in genes involved in accurate chromosome segregation can lead to the formation of MN.

9.4 MN and Genetic Polymorphisms in Metabolic Activation, Detoxification of Genotoxicants and Folate Metabolism

Additional genetic polymorphisms (or mutations) that can influence MN frequencies are those in genes responsible for the metabolic activation and detoxification of clastogens, and in folate metabolism.

9.4.1 Metabolic Activation and Detoxification of Genotoxicants

The polymorphic P450 (CYP) enzyme superfamily is the most important system involved in the biotransformation of many endogenous and exogenous substances including drugs, toxins and carcinogens. Genes coding for *CYP1A1*, *CYP1A2*, *CYP1B1* and *CYP2E1* are among those responsible for the biotransformation of chemicals, especially for the metabolic activation of precarcinogens to genotoxic intermediates. There is evidence of association

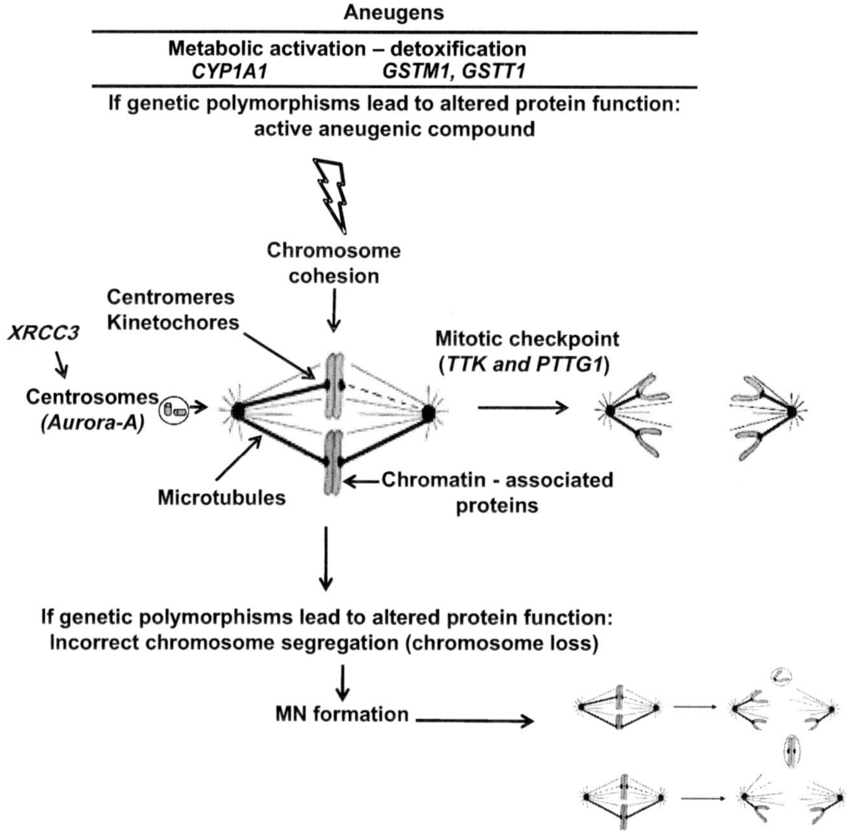

Figure 9.2 Model of how genetic polymorphisms in genes involved in xenobiotic metabolism and accurate chromosome segregation can lead to the formation of MN arising by chromosome loss after exposure to aneugens. The genes mentioned are examples discussed in the text.

between polymorphisms in these genes and cancer susceptibility.[41] The influence of genetic polymorphisms in these genes on MN frequencies has been studied in association with various environmental/occupational exposures.[42] In particular, polymorphisms in *CYP2E1* (c1/c2) showed a strong association with MN induction.[17]

As far as detoxificartion of genotoxicants is concerned, human glutathione *S*-transferases (GSTs) belong to a multi-gene family of four different classes of detoxification isozymes (α, μ, π and θ), which are involved in detoxification of xenobiotics by conjugating a wide range of different chemicals with reduced glutathione (GSH).[43] *GSTM1*, *T1* and *A1* are known to be polymorphic. The so-called "null" *GSTM1* and *GSTT1* genotypes have demonstrated well-established functional relevance, *i.e.* reduced enzyme activity, which in turn seems to denote impaired ability to detoxify carcinogens, a state possibly conferring increased cancer risk.[44] In general, genetic polymorphisms leading to a decreased GST activity have been reported to increase cancer risk, although a protective effect of these polymorphisms for different kinds of cancer has also been described.[45] Some studies on *GSTM1* found higher MN frequencies in positive individuals compared to their null counterparts, while others found the opposite effect and within exposed individuals, controls, or the total population. In a number of studies *GSTT1*-null subjects had higher MN frequencies than their positive counterparts, while in other studies higher MN frequencies were reported for *GSTT1*-positive individuals.[17,42] A pooled analysis by Kirsch-Volders *et al.*[46] of data from seven laboratories performing biomonitoring studies using the CBMN assay revealed that *GSTT1* genotype influenced MN frequencies in an age-dependent way. Lower MN frequencies were found in *GSTT1*-null subjects aged 20, and higher MN frequencies in *GSTT1*-null subjects aged 60, when compared with *GSTT1*-positive genotypes. As far as aneugenicity is concerned, few data are available. Kim *et al.*[47] demonstrated that specific genotypes of *CYP* and *GST* were associated with increased frequencies of aneuploidy in benzene-exposed individuals, suggesting that specific combinations of polymorphic genes might confer a predisposition to aneuploidy.

In general one can expect that individuals with variant genotypes in metabolic activation genes, leading to an increased activity of the proteins and thus activation of genotoxicants, are at higher risk of cancer and of MN induction. For individuals carrying variant genotypes in detoxifying enzymes, on the other hand, a decreased activity of the proteins would be associated with an increased risk of cancer and MN induction. However, controversial data can often be found, and this can be explained partly by the overlapping substrate specificities of different iso-enzymes. Moreover, it should be emphasized that several pathways often act together in the metabolism of mutagens. Furthermore, the induction of enzyme activities in various tissues may not be identical, suggesting that caution is required in reaching conclusions.

9.4.2 Folate Metabolism

Folate deficiency causes chromosome breaks and chromosome loss, and MN formation, via: (i) excessive uracil incorporation into DNA due to insufficient methylation of deoxyuridine monophosphate to thymidine-5-monophosphate and (ii) hypomethylation of cytosine due to inadequate methionine and S-adenosyl methionine synthesis.[48] Many studies have shown an association between folate deficiency and increased DNA damage or MN formation. Two key enzymes in folate metabolism are 5,10-methylenetetrahydrofolate reductase (MTHFR) and methionine synthase (MS).[48] Considering the importance of folate metabolism for the maintenance of genome stability, several epidemiological studies have been undertaken to assess the association between genetic polymorphisms in folate-related genes and cancer risk. The variant $MTHFR^{222}$ and MS^{919} alleles have been associated with an increased risk of various cancers, although a lack of association with cancer risk or a decreased cancer risk have also been described. Many SNPs have been reported in these genes that are capable of influencing MN frequencies.[48] Dhillon *et al.*[49] reported that individuals from a South Australian cohort who carried at least one copy of the rarer G allele for *MTR* (A2756G), or were homozygous for the more common G allele for *RFC* (G80A), had lower MN frequency compared with those with the alternative genotypes for each SNP respectively. Genotype combination analyses showed that MN frequencies were highest in those with the combined *MTR* (2756) AA and *RFC* (80) GA or AA genotype and lowest in those with the combined *MTR* (2756) AG or GG and *RFC* (80) GG genotypes. Polymorphisms in the reduced folate carrier gene (*RFC*; G80A) were significantly associated with the higher MN in Australians.[17] Similarly, MN frequency is increased in 'A' allele carriers of the *MTR* gene (A2756G) polymorphism in non-smokers and smokers in Australian and Japanese populations, respectively.[17] There are two functional polymorphisms in the *MTHFR* gene (C677T and A1298C). The TT genotype in the *C677T MTHFR* polymorphism is significantly associated with the higher MN in coronary artery disease patients.[17] In addition, some studies on *MTHFR*^{677} found higher MN frequencies in individuals carrying the TT variant as compared with the CC or CT genotypes, or the CC genotype alone.[42] However to suggest definitive associations more studies are needed.[17,42]

9.5 Conclusion and Future Perspectives

It is clear that MN induction can be affected by genetic polymorphisms in various genes involved in DNA repair pathways, xenobiotic metabolism and folate metabolism and this is most often the case for MN induced by clastogenic compounds. The association between genetic polymorphisms of genes playing a role in the different mechanisms of correct chromosome segregation, however, is poorly known, and the link with MN formation has not yet been evaluated. Future studies are required to increase understanding of MN formation driven by genetic polymorphisms, especially for aneuploidy-related genetic polymorphisms. Therefore studies should also include additional endpoints, such as MN-FISH (micronucleus assay combined with fluorescence *in situ* hydridisation) with pancentromeric probes, nucleoplasmic bridges and nuclear buds. In addition, large-scale studies including age, gender and lifestyle factors should be considered in future studies. Moreover, the combination of genetic variants is an important issue, and therefore an understanding of gene–gene interactions is required.

Acknowledgements

This work was supported by the EU research programmes "The Detection and Hazard Evaluation of Aneugenic Chemicals" (ENV4-CT97-0471), "Protection of the European Population from Aneugenic Chemicals (PEPFAC)" (QLK4-CT-2000-00058), Cancer Risk Biomarkers (Contract no. QLK4-2000-00628),

ECNIS (Environmental Cancer Risk, Nutrition and Individual Susceptibility), a network of excellence operating within the European Union 6th Framework Program, Priority 5: "Food Quality and Safety" (Contract No 513943) and by the EU Integrated Project NewGeneris ("Newborns and Genotoxic exposure risk")', 6th Framework Programme, Priority 5: Food Quality and Safety (Contract no. FOOD-CT-2005-016320). NewGeneris is the acronym of the project, the Belgian Science Policy [contractnumber SD/HE/02B]; OZR-VUB.

References

1. W. W. Au, P. Navasumrit and M. Ruchirawat, Use of biomarkers to characterize functions of polymorphic DNA repair genotypes, *Int. J. Hyg. Environ. Health*, 2004, **207**, 301–313.
2. M. C. Mahoney, Genetic polymorphisms and disease prevention, *Pediatr. Blood Cancer*, 2007, **48**, 742–747.
3. R. Sachidanandam, D. Weissman, S. C. Schmidt, J. M. Kakol, L. D. Stein, G. Marth, S. Sherry, J. C. Mullikin, B. J. Mortimore and D. L. Willey, A map of human genome sequence variation containing 1.42 million single nucleotide polymorphisms, *Nature*, 2001, **409**, 928–933.
4. E. S. Lander, L. M. Linton, B. Birren, C. Nusbaum, M. C. Zody, J. Baldwin, K. Devon, K. Dewar, M. Doyle and W. FitzHugh, Initial sequencing and analysis of the human genome, *Nature*, 2001, **409**, 860–921.
5. G. Orphanides and I. Kimber, Toxicogenetics: applications and opportunities, *Toxicol. Sci.*, 2003, **75**, 1–6.
6. S. Bonassi, D. Ugolini, M. Kirsch-Volders, U. Strömberg, R. Vermeulen and J. D. Tucker, Human population studies with cytogenetic biomarkers: review of the literature and future prospectives, *Environ. Mol. Mutagen.*, 2005, **45**, 258–270.
7. M. Fenech, Cytokinesis-block micronucleus cytome assay, *Nat. Protoc.*, 2007, **2**, 1084–1104.
8. M. Kirsch-Volders, G. Plas, A. Elhajouji, M. L. Gonzalez, K. Vande Loock and I. Decordier, The *in vitro* MN assay in 2010: origin and fate, biological significance, protocols, high throughput methodologies and toxicological relevance, *Arch. Toxicol.*, 2011, in press.
9. R. Mateuca, N. Lombaert, P. V. Aka, I. Decordier and M. Kirsch-Volders, Chromosomal changes: induction, detection methods and applicability in human biomonitoring, *Biochimie*, 2006, **88**, 1515–1531.
10. R. A. Mateuca, M. Roelants, G. Iarmarcovai, P. V. Aka, L. Godderis, A. Tremp, S. Bonassi, M. Fenech, J. L. Bergé-Lefranc and M. Kirsch-Volders, hOGG1(326), XRCC1(399) and XRCC3(241) polymorphisms influence micronucleus frequencies in human lymphocytes *in vivo*, *Mutagenesis*, 2008, **23**, 35–41.
11. S. Marsin, A. E.Vidal, M. Sossou, J. Menissier-de Murcia, F. Le Page, S. Boiteux, G. de Murcia and J. P.Radicella, Role of XRCC1 in the

coordination and stimulation of oxidative DNA damage repair initiated by the DNA glycosylase hOGG1, *J. Biol. Chem.*, 2003, **278**, 44068–44074.
12. K. W. Caldecott, XRCC1 and DNA strand break repair, *DNA Rep.*, 2003, **2**, 955–969.
13. L. Godderis, M. De Boeck, V. Haufroid, M. Emmery, R. Mateuca, S. Gardinal, M. Kirsch-Volders, H. Veulemans and D. Lison, Influence of genetic polymorphisms on biomarkers of exposure and genotoxic effects in styrene-exposed workers, *Environ. Mol. Mutagen.*, 2004, **44**, 293–303.
14. T. Cornetta, L. Padua, A. Testa, E. Ievoli, F. Festa, G. Tranfo, L. Baccelliere and R. Cozzi, Molecular biomonitoring of a population of nurses handling antineoplastic drugs, *Mutat. Res.*, 2008, **638**, 75–82.
15. Q. Wang, F. Ji, Y. Sun, Y. L. Qiu, W. Wang, F. Wu, W. B. Miao, Y. Li, P. W. Brandt-Rauf and Z. L. Xia, Genetic polymorphisms of XRCC1, HOGG1 and MGMT and micronucleus occurrence in Chinese vinyl chloride-exposed workers, *Carcinogenesis*, 2010, **31**, 1068–1073.
16. F. Ji, W. Wang, Z. L. Xia, Y. J. Zheng, Y. L. Qiu, F. Wu, W. B. Miao, R. F. Jin, J. Qian, L. Jin, Y. L. Zhu and D. C. Christiani, Prevalence and persistence of chromosomal damage and susceptible genotypes of metabolic and DNA repair genes in Chinese vinyl chloride-exposed workers, *Carcinogenesis*, 2010, **31**, 648–653.
17. V. S. Dhillon, P. Thomas, G. Iarmarcovai, M. Kirsch-Volders, S. Bonassi and M. Fenech, Genetic polymorphisms of genes involved in DNA repair and metabolism influence micronucleus frequencies in human peripheral blood lymphocytes, *Mutagenesis*, 2011, **26**, 33–42.
18. J. Cheng, S. Leng, H. Li, C. Huang, Y. Niu, L. Zhang, X. Liang, H. Lin and Y. Zheng, Suboptimal DNA repair capacity predisposes coke-oven workers to accumulate more chromosomal damages in peripheral lymphocytes, *Cancer Epidemiol. Biomarkers Prev.*, 2009, **18**, 987–993.
19. L. H. Thompson and M. G. West, XRCC1 keeps DNA from getting stranded, *Mutat. Res.*, 2000, **459**, 1–18.
20. M. O'Driscoll and J. A. Jeggo, The role of double-strand break repair – insights from human genetics, *Nat. Rev. Genet.*, 2006, **7**, 45–54.
21. G. Matullo, D. Palli, M. Peluso, S. Guarrera, S. Carturan, E. Celentano, V. Krogh, A. Munnia, T. Rumino and S. Polidoro , XRCC1, XRCC3, XPD gene polymorphisms, smoking and (32)P-DNA adducts in a sample of healthy subjects, *Carcinogenesis*, 2001, **22**, 1437–1445.
22. W. W. Au, S. A Salama and C. H. Sierra-Torres, Functional characterization of polymorphisms in DNA repair genes using cytogenetic challenge assays, *Environ. Health. Perspect.*, 2003, **111**, 1843–1850.
23. P. Aka, R. Mateuca, J. P. Buchet, H. Thierens and M. Kirsch-Volders, Are genetic polymorphisms in OGG1, XRCC1 and XRCC3 genes predictive for the DNA strand break repair phenotype and genotoxicity in workers exposed to low dose ionizing radiations? *Mutat. Res.*, 2004, **556**, 169–181.
24. S. Angelini, R. Kumar, F. Carbone, F. Maffei, G. C. Forti, F. S. Violante, V. Lodi, S. Curti, K. Hemminki and P. Hrelia, Micronuclei in humans

induced by exposure to low level of ionizing radiation: influence of polymorphisms in DNA repair genes, *Mutat. Res*., 2005, **570**, 105–117.
25. M. G. Andreassi, I. Foffa, S. Manfredi, N. SBotto, A. Cioppa and E. Picano, Genetic polymorphisms in XRCC1, OGG1, APE1 and XRCC3 DNA repair genes, ionizing radiation exposure and chromosomal DNA damage in interventional cardiologists, *Mutat. Res.*, 2009, **666**, 57–63.
26. I. Decordier, K. De Bont, K. De Bock, R. Mateuca, M. Roelants, R. Ciardelli, D. Haumont, L. E. Knudsen and M. Kirsch-Volders, Genetic susceptibility of newborn daughters to oxidative stress, *Toxicol. Lett.*, 2007, **172**, 68–84.
27. R. Mateuca, P. V. Aka, M. De Boeck, R. Hauspie, M. Kirsch-Volders and D. Lison, Influence of hOGG1, XRCC1 and XRCC3 genotypes on biomarkers of genotoxicity in workers exposed to cobalt or hard metal dusts, *Toxicol Lett.*, 2005, **156**, 277–288.
28. M. Kirsch-Volders and M. Fenech, Inclusion of micronuclei in non-divided mononuclear lymphocytes and necrosis/apoptosis may provide a more comprehensive cytokinesis block micronucleus assay for biomonitoring purposes, *Mutagenesis*, 2001, **16**, 51–58.
29. I. Decordier, R. Mateuca and M. Kirsch-Volders, Micronucleus assay and labeling of centromeres with FISH technique, *Methods Mol. Biol.*, 2011, **691**, 115–136.
30. Y. H. Cho, Y. J. Kim, Y. S. An, H. D. Woo, S. Y. Choi, C. M. Kang and H. W. Chung, Micronucleus-centromere assay and DNA repair gene polymorphism in lymphocytes of industrial radiographers, *Mutat. Res.*, 2009, **680**, 17–24.
31. G. Iarmarcovai, I. Sari-Minodier, T. Orsière, M. De Méo, P. Gallice, C. Bideau, D. Iniesta, J. Pompili, J. L. Bergé-Lefranc and A Botta, A combined analysis of XRCC1, XRCC3, GSTM1 and GSTT1 polymorphisms and centromere content of micronuclei in welders, *Mutagenesis*, 2006, **2**, 159–165.
32. E. L. Goode, C. M. Ulrich and J. D. Potter, Polymorphisms in DNA repair genes and associations with cancer risk, *Cancer Epidemiol. Biomarkers Prev.*, 2002, **11**, 1513–1530.
33. S. Sen, Aneuploidy and cancer, *Curr. Opin. Oncol.*, 2000, **12**, 82–88.
34. V. M. Draviam, S. Xie and P. K. Sorger, Chromosome segregation and genomic stability, *Curr. Opin. Genet. Dev.*, 2004, **14**, 120–125.
35. Y. L. Lo, J. C. Yu, S. T. Chen, H. C. Yang, C. S. Fann, Y. C. Mau and C. Y. Shen, Breast cancer risk associated with genotypic polymorphism of the mitosis-regulating gene Aurora-A/STK15/BTAK, *Int. J. Cancer*, 2005, **115**, 276–283.
36. Y. L. Lo, J. C. Yu, S. T. Chen, G. C. Hsu, Y. C. Mau, S. L. Yang, P. E. Wu and C. Y. Shen, Breast cancer risk associated with genotypic polymorphism of the mitotic checkpoint genes: a multigenic study on cancer susceptibility, *Carcinogenesis*, 2007, **28**, 1079–1086.
37. J. E. Olson, X. Wang, V. S. Pankratz, Z. S. Fredericksen, C. M. Vachon, R. A. Vierkant, J. R. Cerhan and F. J. Couch, Centrosome-related genes,

genetic variation, and risk of breast cancer, *Breast Cancer Res. Treat.*, 2011, **125**, 221–228.
38. T. Yoshihara, M. Ishida, A. Kinomura, M. Katsura, T. Tsuruga, S. Tashiro, T. Asahara and K. Miyagawa, XRCC3 deficiency results in a defect in recombination and increased endoreduplication in human cells, *EMBO J.*, 2004, **23**, 670–680.
39. A. R. Lindh, S. Rafii, N. Schultz, A. Cox and T. Helleday, Mitotic defects in XRCC3 variants T241M and D213N and their relation to cancer susceptibility, *Hum. Mol. Genet.*, 2006, **15**, 1217–1224.
40. D. Cimini and F. Degrassi, Aneuploidy: a matter of bad connections, *Trends Cell Biol.*, 2005, **15**, 442–451.
41. N. Bozina, V. Bradamante and M. Lovrić, Genetic polymorphism of metabolic enzymes P450 (CYP) as a susceptibility factor for drug response, toxicity, and cancer risk, *Arh. Hig. Rada. Toksikol.*, 2009, **60**, 217–242.
42. G. Iarmarcovai, S. Bonassi, A. Botta, R. A. Baan and T. Orsière, Genetic polymorphisms and micronucleus formation: a review of the literature, *Mutat. Res.*, 2008, **658**, 215–233.
43. J. D. Hayes and D. J. Pulford, The glutathione S-transferase supergene family: regulation of GST and the contribution of the isoenzymes to cancer chemoprotection and drug resistance, *Crit. Rev. Biochem. Mol. Biol.*, 1995, **30**, 445–600.
44. K. P. Economopoulos and T. N. Sergentanis, GSTM1, GSTT1, GSTP1, GSTA1 and colorectal cancer risk: a comprehensive meta-analysis, *Eur. J. Cancer*, 2010, **46**, 1617–1631.
45. G. Di Pietro, L. A. Magno and F. Rios-Santos, Glutathione S-transferases: an overview in cancer research, *Expert. Opin. Drug Metab. Toxicol.*, **6**, 153–170.
46. M. Kirsch-Volders, R. A. Mateuca, M. Roelants, A. Tremp, E. Zeiger, S. Bonassi, N. Holland, W. P. Chang, P. V. Aka, M. Deboeck, L. Godderis, V. Haufroid, H. Ishikawa, B. Laffon, R. Marcos, L. Migliore, H. Norppa, J. P. Teixeira, A. Zijno and M. Fenech, The effects of GSTM1 and GSTT1 polymorphisms on micronucleus frequencies in human lymphocytes *in vivo*, *Cancer Epidemiol. Biomarkers Prev.*, 2006, **15**, 1038–1042.
47. S. Y. Kim, Chromosomal aberrations in workers exposed to low levels of benzene: association with genetic polymorphisms, *Pharmacogenetics*, 2004, **14**, 453–463.
48. J. Y. Fang and S. D. Xiao, Folic acid, polymorphism of methyl-group metabolism genes, and DNA methylation in relation to GI carcinogenesis, *J. Gastroenterol.*, 2003, **38**, 821–829.
49. V. Dhillon, P. Thomas and M. Fenech, Effect of common polymorphisms in folate uptake and metabolism genes on frequency of micronucleated lymphocytes in a South Australian cohort, *Mutat. Res.*, 2009, **665**, 1–6.

CHAPTER 10

Biomarkers of Effects on Hormone Functions

MANDANA GHISARI, TANJA KRÜGER, MANHAI LONG AND EVA CECILIE BONEFELD-JØRGENSEN*

Centre of Arctic Environmental Medicine, School of Public Health, Bartholins Allé 2, 8000 Aarhus C, University of Aarhus, Denmark

10.1 Introduction

Endocrine-disrupting chemicals (EDCs) are compounds that either mimic or block endogenous hormones and can disrupt the normal function of the body. The European Commission has defined an EDC as "an exogenous substance or mixture, that alters the function(s) of the endocrine system, and consequently causes adverse health effects in an intact organism or its progeny or (sub)-population". Growing evidence shows that EDCs may also modulate the activity and/or expression of steroidogenic enzymes. These are expressed not only in the adrenal glands and gonads, but also in many tissues that have the ability to convert circulating precursors into active hormones.

Many EDCs are known to act as agonist/antagonists of estrogen (ER), androgen (AR) and aryl hydrocarbon (AhR) receptors, to interfere with thyroid hormone (TH) function and to inhibit the aromatase enzyme that converts testosterone to estradiol (E2; reviewed in ref. 1). This is of particular concern for the developing organism, because it is highly sensitive to hormonal changes and EDC exposure, which can result in changes that are permanent.[2]

Issues in Toxicology No. 10
Biomarkers and Human Biomonitoring, Volume 2: Selected Biomarkers of Current Interest
Edited by Lisbeth E. Knudsen and Domenico Franco Merlo

Published by the Royal Society of Chemistry, www.rsc.org

The broad categories of human health effects that may be linked to exposure to environmental contaminants include: cancer, birth defects, decreased fertility, altered sex hormone balance, immune system defects, neurological effects such as reduced IQ and behavioral abnormalities, altered metabolism, and specific organ dysfunctions.[3]

The EDCs may occur naturally, for example phytoestrogens, or may be industrial chemicals such as bisphenol A (BPA) and other phenols and plasticizers such as the phthalates and adipates commonly used in the plastic industry. The EDCs also include the persistent organic pollutants (POPs) such as polychlorinated biphenyls (PCBs), polychlorinated dibenzo-*p*-dioxins/furans (PCDDs/PCDFs), organochlorine pesticides (OCPs), brominated flame retardents (BFRs) and perfluorinated compounds.

We are exposed concurrently to a complex mixture of chemicals from various sources and the toxicological assessment of EDCs in human is complicated. Toxicological studies have shown that individual POPs have very different biological effects and potentials; many of the bioaccumulated POPs are estrogenic while others are antiestrogenic and antiandrogenic and some have dioxin-like potentials.[3–9] As a result of these different and often opposed biological effects and potentials of the POPs, it is very difficult, if not impossible, to predict a given biological effect of the very complex mixtures of POPs that actually exist in the human body. Furthermore, additive enhancement of hormone actions has been reported *in vitro* for xenoestrogen mixtures[10,11] and recently *in vivo* for antiandrogens.[12] Therefore, the assessment of the integrated biological effect of the actual chemical mixture in human blood is important and cell systems have been introduced recently to enable the assessment *ex vivo* of the integrated effect of xenobiotic compounds in human adipose tissue[13–15] or in human serum.[16–20]

Biomonitoring is the assessment of internal doses of EDCs and has been used for decades to provide information about exposures to chemicals, giving information about a person's "body burden" of EDCs.

Human biomonitoring (HBM) exploits the development and characterization of biomarkers, *i.e.*, "cellular, biochemical or molecular alterations that are measurable in biological media such as human tissues, cells, or fluids", to measure direct and indirect human exposure to EDCs. In fact, biomonitoring can integrate environmental exposure, effects, individual sensitivity and health risks, making it a useful instrument for risk assessment. In particular, biomonitoring can identify trends and changes in current exposure patterns as well as establishing the distribution of exposure among the general population, identifying sub-groups and/or populations with higher exposure levels.

In this chapter, we will give an overview of our *in vitro* work performed on single EDCs and their mixtures in five key bioassays for effects on ER, AR, and AhR transactivation, TH action and aromatase activity. Furthermore, *ex vivo* results for the integrated potential of extracted serum POPs to affect the function of ER and AR, as well as AhR-mediated dioxin-like transactivity, as biomarkers for effects will be reviewed.

10.2 Methods

10.2.1 *In vitro*

10.2.1.1 Transactivation Assays

The transactivation assays are based on genetically modified cell lines in which specific receptor responsive DNA elements are linked to a firefly luciferase reporter gene that transcribes to the easily measurable luciferase protein (see Figure 10.1). Basically, these transactivation assays measure receptor-induced gene expression, which gives information about the expected biological response to chemicals in humans. For example, measurement of elevated or reduced gene expression indicates whether specific chemicals would exert agonistic or antagonistic effects on the receptor function. Several transactivation assays have been utilized with respect to endocrine disruption of estrogen, androgen, and aryl hydrocarbon receptors: the ER transactivation assay, AR transactivation assay and AhR transactivation assay for detection of dioxin-like compounds. These assays have been used to study receptor-mediated transactivities

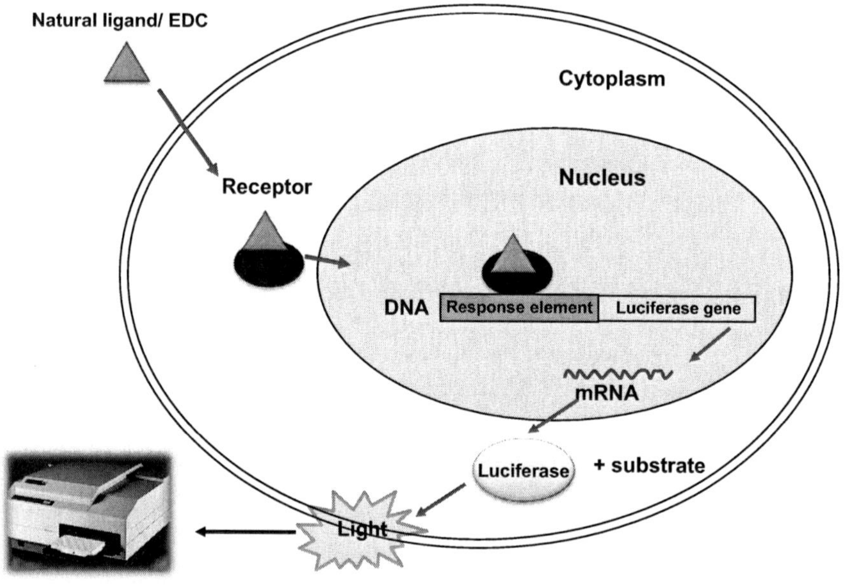

Figure 10.1 The principle of the transactivation assays. When a compound eliciting estrogenic/androgenic/dioxin-like transactivity interacts with the receptor, the ligand–receptor complex will be activated and translocate to the nucleus. Subsequently, the ligand–receptor complex will bind to specific response elements of the DNA, thus activating luciferase gene expression. The resulting mRNA translates into luciferase enzyme, which catalyzes a bioluminescent reaction, using luciferin as a substrate. The amount of light is proportional to the estrogenic/androgenic dioxin-like transactivity of the compounds. The luciferase data are given as relative light units (RLU), and subsequently this is corrected to the protein concentration of the exposed cells.

of individual chemicals and mixtures[4,5,8,21,22] as well as for epidemiological purposes.[16–18,23,24]

ER transactivation assay: The ability of the xenobiotics to affect ER function was assessed using the ER transactivation assay. For this assay the MVLN cell line, derived from the human breast carcinoma MCF-7 cell line, was used. MVLN cells express both ER subtypes, ERα and ERβ, and furthermore they are stably transfected with an ERE-luciferase reporter vector.[16,21] The samples were tested alone or by co-treatment with the high affinity ligand E2 to assess the direct interference with the receptor (agonism) or the competition (antagonism) with the natural hormone, respectively.

AR transactivation assay: The ability of the xenobiotics to affect AR function was assessed by the AR transactivation assay. This assay uses the Chinese hamster ovary cell line CHO-K1 transiently co-transfected with a MMTV-luciferase reporter vector and the AR expression plasmid pSVAR0.[17] The effects of the xenobiotics were determined alone or in the presence of the synthetic AR agonist R1881 to assess antagonistic transactivities.

AhR transactivation assay: The effects of xenobiotics on the transactivation of the AhR were analyzed in the mouse hepatoma Hepa1.12cR cell line, stably transfected with the PAH/HAH-inducible luciferase expression vector containing dioxin responsive elements, and induction of luciferase occurs in a time-, dose- and AhR-dependent manner.[18] The samples were tested alone or in co-treatment with TCDD.

10.2.1.2 T-screen

The interference of xenobiotics with the action of TH was studied using the T-screen assay, which is based on the TH-dependent cell growth of a rat pituitary tumor cell line, GH3. The GH3 cells express both endogenous TRs and ERs and can synthesize and secrete prolactin and growth hormone, in response to estrogen and TH, respectively. GH3 cells respond to physiological levels of triiodothyronine (T3), resulting in increased GH production and cell proliferation[25] and have been used for *in vitro* detection of agonistic and antagonistic properties of compounds at the level of the TR.[26,27] The T-screen was carried out using rat essentially as described previously.[27] To assess the agonistic and antagonistic effects the xenobiotics were tested in the absence or presence of T3.

10.2.1.3 Aromatase assay

Aromatase cytochrome P450 is the key enzyme in the biosynthesis of estrogens from cholesterol, because it catalyzes the final, rate-limiting step in which androgens are converted to estrogens. The conversion proceeds via three successive oxidation steps, and this is utilized in the aromatase assay measuring the release of tritium-labeled water (Figure 10.2).

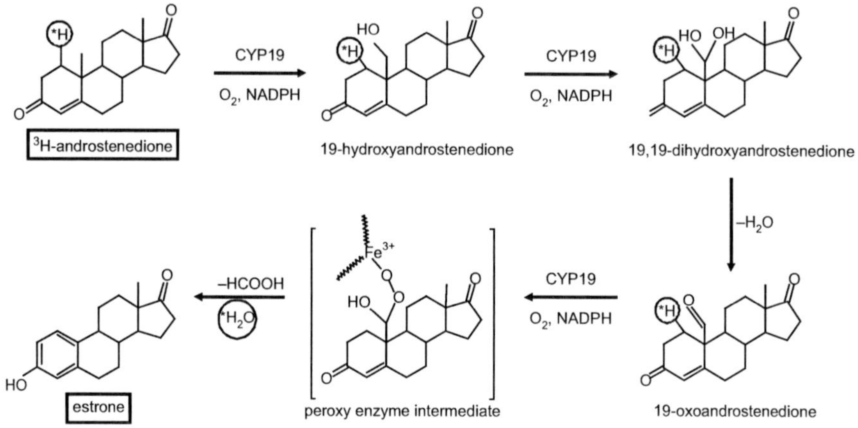

Figure 10.2 Conversion of androstenedione to estrone by the aromatase enzyme.

The effects on aromatase enzyme activity were assessed using the human choriocarcinoma JEG-3 cell line, derived from malignant placental tissues. JEG-3 cells endogenously express aromatase activity at detectable levels. The catalytic aromatase activity was determined by the release of tritium as described previously,[28] with minor modifications. The aromatase activity was determined after 18 hours of cell growth followed by 2 hours of co-exposure to 0.2 μCi [^{3}H]-4-androstene-3,17-dione (4-AD) and 10 nM unlabeled 4-AD and test compounds. The final results were normalized to cell protein content.[22]

10.2.1.4 Principle and Practice in Mixture Analyses

Humans are exposed to a complex mixture of chemicals. Therefore, assessment of the combined effect of mixtures is very important. A method based on the principle of concentration addition (CA) has been shown to be a valid tool for assessing mixture effects of similarly and dissimilarly acting EDCs *in vitro*.[7,10,29] The CA model assumes that one chemical can be replaced totally or partly by another in the mixture, without changing the overall combined effect, and that they act through a similar mechanism.[30] By applying the principle of concentration addition, the concentration of the compounds in a mixture, at an observed mixture effect, can be predicted using the concentration–response data for each compound alone.[8,31] Knowledge of the ratio of the compounds in the mixture is a prerequisite for using this method.

10.2.2 *Ex Vivo*

To obtain the serum fraction containing the actual mixture of bio-accumulated lipophilic POPs for determination of ER and AR transactivity measurement, a solid phase extraction (SPE) and high performance liquid chromatography (HPLC) fractionation was performed on 3.6 ml serum[19] (Figure 10.3). The first

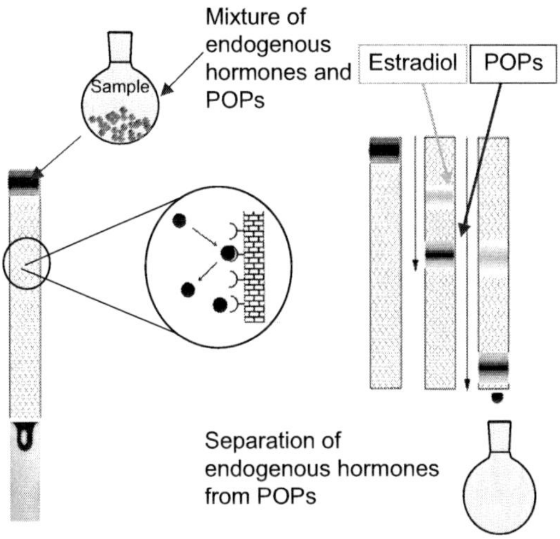

Figure 10.3 The principle of high performance liquid chromatography (HPLC) fractionation. The serum sample, illustrated here by a mixture of a hormones and POPs, is loaded on the HPLC column. Interactions between the column filling and different compounds in the sample are not equally strong, resulting in some compounds moving faster than others. The fastest moving compounds, such as the POPs, will elute early and can be collected by a fraction collector, before any of the endogenous hormones are eluted.

fraction (F1: 0.00–5.30 min.) was defined to include most lipophilic POPs and was free of endogenous hormones.[19]

The extraction of the serum samples to obtain the fraction containing lipophilic POPs for AhR transactivity measurements was performed at a certified laboratory, Le Centre de Toxicologie, Sainte Foy, Quebec, Canada.[32]

The serum extracts were stored at –80°C and on the day of analysis the extracts were thawed, processed and analyzed as described for the ER, AR and AhR transactivation assays.[16–18]

10.3 Results

10.3.1 *In vitro*

In the following we will summarize the relative endocrine disrupting (ED) potencies of a range of EDCs that were analyzed in five key *in vitro* bioassays for ED effects including ER, AR, and AhR transactivation, TH action and aromatase activity.

10.3.1.1 PCBs

Research suggests that mechanisms of PCB toxicity are multifactorial, and they appear to involve both AhR and other hormone receptor mechanisms.

The most frequently detected di-ortho and higher substituted PCBs in humans are CB138, CB153, and CB180,[33] which have been shown to contribute about 50–80% of the total PCB body burden in for example the Swedish general population.[34] These three PCBs, individually and in a mixture, as well as other non- and di-ortho substituted PCBs, were shown *in vitro* to interfere with cell proliferation as well as the function of the ER and AR.[4,19,35] Similarly, some hydroxylated PCBs (OH-PCBs), major metabolites of PCBs, exert low estrogenic and/or antiestrogenic effects.[20,36]

Non- or mono-ortho PCBs are ligands to the AhR, whereas ortho-substituted PCBs are either very weak ligands or do not bind to the AhR.[37] However, the higher chlorinated PCBs have the potential to inhibit the function of the AhR.[38]

The OH-PCBs have a high degree of structural resemblance to THs and the three OH-PCBs (OH-CB 121, OH-CB 69, OH-CB 106) were shown to be TH-like by stimulating the proliferation of GH3 cells in the T-screen assay.[27] These OH-PCBs were reported previously to bind to the TR and TH transport protein TTR.[39]

10.3.1.2 Pesticides

Pesticides comprise a large number of different substances with dissimilar structures and diverse toxicity. Several pesticides exert ED activities through interaction with *e.g.* the hormone receptors. The OCP *o,p'*-DDT possess estrogenic activities and its metabolite *p,p'*-DDE has antiandrogenic activities,[40] whereas toxaphene is reported to be antiestrogenic.[5] Andersen *et al.*[6] have reported that several currently used pesticides, such as dieldrin, endosulfan, methiocarb, fenarimol, chlorpyrifos, deltamethrin, and tolclofos-methyl, possess estrogenic activity on the basis of the cell proliferation assay and ER transactivation assay using MCF-7 human breast cancer cells. In the same study, dieldrin, endosulfan, methiocarb, prochloraz and fenarimol possessed antiandrogenic activity on the basis of the AR transactivation assay using CHO-K1 cells. Furthermore, fenarimol and prochloraz were potent aromatase inhibitors while endosulfan was a weak inhibitor.[6] In a subsequent study, prochloraz, iprodion and chlorpyrifos were shown to exert dioxin-like AhR agonistic effects.[9] In the T-screen assay, chlorpyrifos weakly stimulated the proliferation of GH3 cells, whereas prochloraz and iprodion inhibited the T3-induced proliferation.[27]

Our results show the diversity of the endocrine effects and the potential of pesticides to affect several endocrine and cellular pathways.

10.3.1.3 Phthalates and Phenols

As part of the EU-sponsored research project ENDOMET, the potential of several plasticizers widely used in industry and present in the environment were assessed *in vitro* for effects on the following receptors, ER, AR, AhR, and on T3-dependent growth of rat pituitary GH3 cells. The plasticizers investigated were: BPA and bisphenol-A dimethacrylate (BPA-DM), alkyl phenols [n-octyl (OP), n-nonyl (nNP), tert-octyl (tOP)], bis-ethylhexyladipate (DEHA), phthalates [dibutyl (DBP), bis-ethylhexyl (DEHP), di-isononyl (DINP), di-isodecyl (DIDP), dioctyl (DOP), benzylbutyl (BBP)] and 2-phenylphenol (2-PP), 4-chloro-3-methyl phenol (CMP), resorcinol and 2,4-dichlorophenol (2,4-DCP). The combined effect of an equipotent mixture of BPA, nNP, BBP, CMP, resorcinol and tOP was also assessed in these assays.

The BPA, BPA-DM, tOP, OP, nNP, BBP and DBP elicited ER-mediated luciferase activity as determined by the ER transactivation assay.[22,31] Using the concentration-addition model, the observed effect concentrations of the mixture were equal to the predicted concentrations, suggesting additivity.

None of the tested plasticizers showed agonistic activities in the AR transactivation assay, whereas, in the presence of the synthetic androgen R1881, several compounds elicited AR antagonistic effects (BPA, BPA-DM, nNP, CMP, Res, tOP, 2-PP, DCP and BBP).[8] The observed mixture effects antagonized the R1881-induced AR response. The predicted effect concentration was equal to the observed effect, again suggesting additivity.

The plasticizers nNP, DBP, DEHP and DIDP elicited weak agonistic AhR transactivity, whereas BPA and CMP inhibited the TCDD-induced AhR transactivity.[8] The mixture composed of six compounds, of which only nNP had a weak agonistic AhR potential, weakly induced the AhR transactivity when compared with the individual compounds. This suggests that the non-AhR active compounds can act together with a weak AhR agonist to increase the AhR transactivity. In the presence of TCDD, the mixture inhibited the TCDD-induced AhR transactivity in a dose-dependent manner.

Generally, the plasticizers tested elicited TH-like potential inducing GH3 cell proliferation, except for nNP, DINP and 2-PP. The nNP inhibited T3-induced GH3 cell growth,[31] and were also reported to antagonize T3-induced luciferase activity.[41]

The TH-like effects of the mixture were less than additive in the absence of T3, indicating an antagonistic effect (interaction of compounds) in GH3 cells. This deviation from additivity might be due to antagonizing effects of NP in the mixture, which potentially can obscure the overall additive effect of the other mixture components.

Of the tested compounds, BPA, nNP, OP, BBP, DEHA, CMP, DBP, DOP and 2,4-DCP were inhibitors of aromatase activity.[22,42]

In summary, based on our studies,[8,22,31] the plasticizers tested have the ability to act via more than one mechanism and this might enhance the biological effect in the intact organism, because the final response is likely to be determined by the interaction of all pathways implicated.

10.3.2 *Ex Vivo*

Laboratory studies on the effects of single chemicals or chemical mixtures in cell cultures and laboratory animals cannot fully elucidate the human health risks. Integration of epidemiological and biomarker studies on humans from exposed populations is needed in order to obtain information about the real health risks resulting from exposure of the body to accumulated mixtures of contaminants. The burden of POPs in Arctic people has been monitored since 1991, and a program for measuring the potential biological effects of these contaminants has been established (Arctic Monitoring and Assessment Programme, AMAP). As part of the AMAP program, the levels of 14 different PCBs, 10 pesticides, selenium, lead and mercury were determined for different Greenlandic districts. Furthermore, to assess the potential biological effects of these contaminants the integrated mixture effects on the function of ER and AR, as well as AhR mediated dioxin-like transactivity, were measured.

10.3.2.1 Biomarkers for POP and Metal Exposures

Figure 10.4 shows Greenland with geographically located columns showing the relative total sum of POP exposure biomarkers, and within each column is shown the relative profile of each POP group determined in human serum, typically including 50 men and 50 women from each district.[43,44]

Both the sum of PCBs and the sum of OCPs are highest in the East Greenlandic districts (EG) (Scoresbysund followed by Tasiilaq), the Thule district in Northern Greenland (NG) and Narsaq in South Greenland (SG). For Qeqertarsuaq (Godhavn, Disco Bay) and Nuuk (West Greenland, WG) the sum of PCBs and the sum of OCPs are very similar. However, it should be noted that the data for Nuuk men are atypical because they were older and had a higher dietary intake of seabirds and, therefore, higher serum POP concentrations. In general the POP levels in Nuuk and Sisimiut, the two most Westernized cities in terms of diet, are at the same level and lower than POP levels in the other Greenlandic districts. What also should be noted is that selenium and mercury serum concentrations were found at the highest level in the Thule area. With respect to serum lead, the highest levels were found in Scoresbysund.

10.3.2.2 Biomarkers for Receptor Effects

As a biomarker for effect, we determined the integrated potential of the serum POP mixture (serum extracts) to affect the function of ER and AR, as well as AhR-mediated dioxin-like transactivity. Currently, the serum POP-related effects on receptor transactivities have been determined for Greenlandic Inuit in Nuuk, Sisimiut, Tasiilaq and Qaanaaq,[23,24] whereas measurement of the effect biomarkers for Scoresbysund (EG), Godhavn (WG) and Narsaq (SG) are in progress.

Figure 10.5 shows the serum POP-related receptor transactivities for Nuuk (WG), Sisimiut (WG), Tasiilaq (EG) and Qaanaaq (NG). For ER transactivity the serum POPs elicited in general an antagonistic effect (21–81% of the samples), with antagonistic activity in the order: Tasiilaq > Sisimiut > Qaanaaq > Nuuk. Few of the serum POP extracts elicited agonistic ER

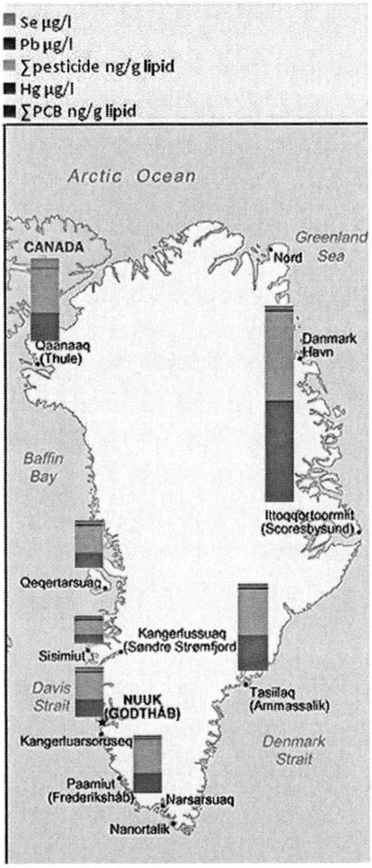

Figure 10.4 Human POP biomarkers for exposure in Greenland. The columns illustrate the relative ∑ POPs and the different POP profiles and metal concentrations within the columns in the different Greenlandic districts for the period 2000–2006.

transactivity ($<6\%$, Qaanaaq $>14\%$). There were observed differences both among districts and between sexes, which in turn was district dependent.[23]

Preliminary data for serum POP-related ER transactivity for Scoresbysund, Godhavn and Narsaq show a trend of higher frequency of serum samples from East Greenland (Scorebysund) eliciting antiestrogenic activity compared with West Greenland (Godhavn and Narsaq) (EC Bonefeld-Jørgensen; unpublished data, 2011). The trend was similar for men and women. These data support observations on the antiestrogenic potentials of POPs because the highest ∑POP concentrations were found in Scorebysund and Tasiilaq.

In addition, for the AR transactivity a district-dependent difference in the serum POP effect was observed with the following order of medians: Nuuk > Sisimiut > Qaanaaq. In general the AR transactivity was higher for men than for women. For men from Nuuk and Sisimiut agonistic AR transactivity was observed.[23] In contrast, a higher frequency of serum extracts elicited anti-androgenic activity for women in Nuuk, Sisimiut, and for both men and women in Qaanaaq.[23] Preliminary data for serum POP related AR transactivity for Scoresbysund, Godhavn and Narsaq shows predominantly antiandrogenic activity (EC Bonefeld-Jørgensen; unpubl.data, 2011), supporting the anti-androgenic potentials of most POPs.

The POP related dioxin-like serum activities are given by TCDD toxicological equivalence (AhR-TEQ). The following order of AhR-TEQ medians was observed: Tasiilaq ≥ Sisimiut ≥ Nuuk > Qaanaaq. The frequency of serum extracts with agonistic AhR transactivity was 71–100% and similar levels were found in women and men.[24] Preliminary data for Scoresbysund, Godhavn and Narsaq also show a high frequency of serum extracts with agonistics AhR transactivity (EC Bonefeld-Jørgensen; unpublished data, 2011).

Comparing the effects of the serum POP mixtures on the three receptor transactivities (see Figure 10.5) there is a trend towards a relationship between higher AhR-TEQ (dioxin activity) level and lower ER transactivity, which supports the reported antiestrogen activity of dioxins.

The study was also intended to evaluate whether the transactivity was associated with serum POP concentrations (14 PCBs and 10 OCPs) and/or

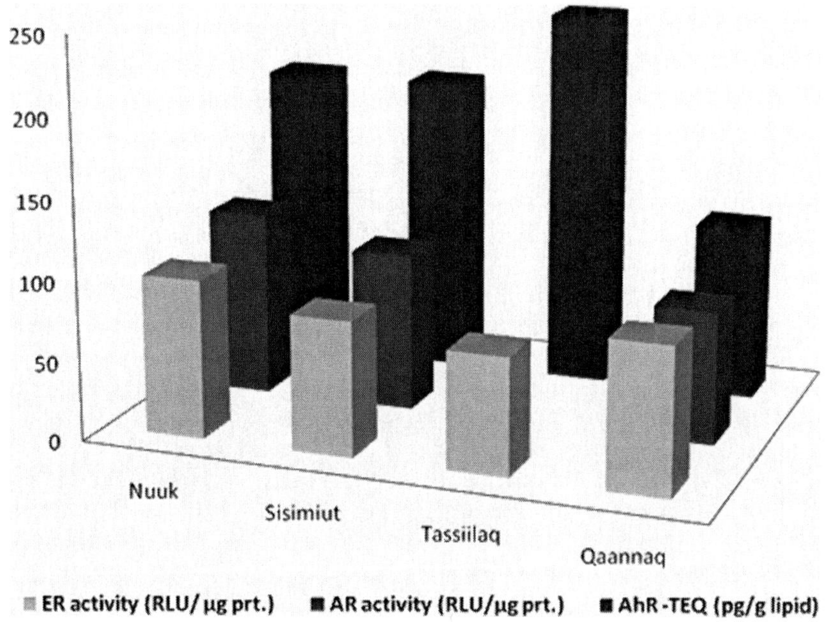

Figure 10.5 Diagram for serum POP-related receptor transactivities in Greenlandic Inuit. ER, estrogen receptor; AR, androgen receptor; AhR, aryl hydrocarbon receptor; RLU/µg prt; relative light units per microgram cell protein. From ref. 44.

lifestyle factors. The ER serum transactivities correlated negatively with the POPs for the combined female data, while the AR and AhR transactivity were negatively correlated with the levels of POPs for the combined male data.

Associations between the ER, AR and AhR transactivities and lifestyle factors such as age, intake of marine food and smoker years were observed, indicating that comparison of different study populations requires the inclusion of age, diet and lifestyle factors. Also, for Scoresbysund, Godhavn and Narsaq preliminary data show correlations between ER and AhR transactivities and POP levels as well as lifestyle factors (EC Bonefeld-Jørgensen; unpublished data, 2011).

Using another clean-up method for serum samples it was reported recently that high levels of PCBs in serum samples from male Slovakians were associated with a decreased ER-mediated transactivity and increased AhR-mediated transactivity.[45]

A few studies using a different setup for measurement of xenoestrogenic activities in adipose tissue from Spanish women and serum samples from pregnant Danish and Faroese women have been reported, although, in contrast to our studies, no correlation between POPs and MCF-7 proliferation were found.[13–15,20] However, recent data showed, for non-pregnant female greenhouse workers in Denmark, that the E2-induced MCF-7 proliferation response was reduced, indicating an antiestrogenic effect of the serum extract containing major xenoestrogens but without pharmaceutical and endogenously produced estrogens.[46]

The project Inuendo, supported by the European Union, aimed to examine whether there is a correlation between human fertility and exposure to POPs, using the POP exposure biomarkers PCB-153 and *p,p′*-DDE. The project involved population groups of fertile men from Europe (Kharkive, Ukraine; Warsaw, Poland; fishermen of the Swedish East Coast) and various Greenlandic districts.[47] A significantly higher serum concentration of the two POP exposure markers was found for Inuit (Figure 10.6). Concerning serum POP-related receptor effect biomarkers, we found a significantly lower ER and AhR transactivity level and a trend towards higher AR transactivity for Greenlandic serum samples in relation to the European samples (Figure 10.6).[16–18]

The POP related serum xenoestrogenic activity of the three European study groups did not differ.[16] The highest frequency of antiandrogenic activity was observed for Ukraine, which also had the highest level of *p,p′*-DDE, a well-known AR antagonist, and a significant negative association between the xenoandrogenic receptor transactivity and *p,p′*-DDE across the combined European study groups was observed.[17] A higher median AhR transactivity level was observed for samples from Poland and Sweden compared to Ukraine.[18]

Unexpectedly, a significantly lower level of DNA damage was found in sperm DNA from Inuit compared with European samples.[48] Further studies are required to identify possible factors involved in the correlations found between POP-related receptor effects and sperm DNA damage.[49,50]

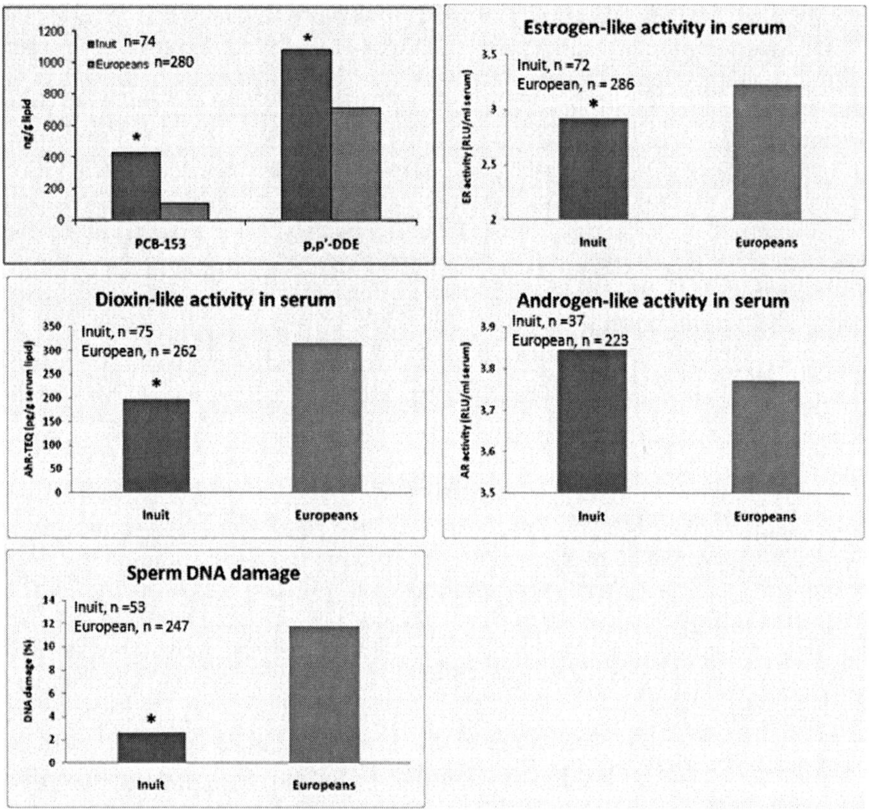

Figure 10.6 Biomarkers for exposure and biomarkers for effects in Inuit and European males. The POP proxy markers (PCB-153 and *p,p′*-DDE) are given in nanograms per gram lipid. Dioxin-like activity is given as aryl hydrocarbon receptor–TCDD toxicological equivalents (AhR-TEQ). ER and AR transactivities are given as relative light units per milliliter serum; $*p < 0.05$, Inuit *vs.* Europeans. From ref. 44.

10.4 Summary, Conclusion and Perspectives

The *in vitro* results showed that EDCs have the ability to act via more than one mechanism and this might enhance the biological effect in the intact organism, because the final response is likely to be determined by the interaction of all pathways implicated. Furthermore, the mixture analyses suggested that the combined effect of all the compounds present in the human body must be taken into consideration for risk assessment.

Biomonitoring studies for exposure biomarkers have shown that there are geographical differences in the levels of bioaccumulated POP in the blood of Europeans and Inuit from different districts. These differences are primarily attributable to diet and lifestyle factors such as smoking, as well as higher levels of traditional Greenlandic diet (*e.g.* seal, whale, polar bear, seabirds) reflected by a higher concentration of POPs in the blood of Inuit compared with Europeans and Inuit with a more westernized diet. The highest POP values were found on the east coast of Greenland, where the inhabitants still rely primarily on traditional Greenlandic foodstuffs.

Biomonitoring studies for receptor effects showed a general correlation between high serum POP concentrations and inhibited ER and AhR transactivity; however, for men in the two West Greenlandic cities Nuuk and Sisimiut a trend towards increased AR transactivity was observed. An inverse trend between dioxin-like AhR transactivity and ER transactivity supports the perception that dioxins exert an antiestrogenic effect. In conclusion, the actual cocktail of serum POPs has endocrine-disrupting potential.

In summary, the xenohormone receptor transactivities can be used as an integrated biomarker of POP exposure and lifestyle characteristics. The data suggest that geographical and sex differences might be caused by the variation in POP profiles in concert with lifestyle and genetic factors.

Biomarkers for POP exposure showed that men from Greenland (Inuit) had significantly higher serum POP levels compared with European men from Sweden, Ukraine and Poland. Unexpectedly, in the same study groups it was observed that Inuit had a lower level of DNA damage in sperm in relation to the Europeans. Further studies are needed to elucidate whether there is a relationship between the impact of serum POPs on the activity of hormone and/or dioxin receptors and the level of sperm DNA damage. However, it is known that selenium and *n*-3 unsaturated fatty acids are important factors in production of semen.

References

1. M. Yang, M. S. Park and H. S. Lee, Endocrine disrupting chemicals: human exposure and health risks, *J. Environ. Sci. Health C Environ. Carcinog. Ecotoxicol. Rev.*, 2006, **24**(2), 183–224.
2. L. J. Guillette, Jr., D. A. Crain, A. A. Rooney and D. B. Pickford, Organization versus activation: the role of endocrine-disrupting contaminants (EDCs) during embryonic development in wildlife, *Environ. Health Perspect.*, 1995, **103**(Suppl. 7), 157–164.
3. E. C. Bonefeld-Jørgensen and P. Ayotte, *Toxicological Properties of POPs and Related Health Effects of Concern for the Arctic Populations, Chapter 6, in AMAP Assessment 2002: Human Health in the Arctic*, ed. AMAP, 2003, Arctic Monitoring and Assessment Programme (AMAP), Oslo, Norway (www.amap.no), pp. 57–73.
4. E. C. Bonefeld-Jorgensen, H. R. Andersen, T. H. Rasmussen and A. M. Vinggaard, Effect of highly bioaccumulated polychlorinated biphenyl congeners on estrogen and androgen receptor activity, *Toxicology*, 2001, **158**(3), 141–153.
5. E. C. Bonefeld-Jorgensen, H. Autrup and J. C. Hansen, Effect of toxaphene on estrogen receptor functions in human breast cancer cells, *Carcinogenesis*, 1997, **18**(8), 1651–1654.

6. H. R. Andersen, A. M. Vinggaard, T. H. Rasmussen, I. M. Gjermandsen and E. C. Bonefeld-Jorgensen, Effects of currently used pesticides in assays for estrogenicity, androgenicity, and aromatase activity in vitro, *Toxicol. Appl. Pharmacol.*, 2002, **179**(1), 1–12.
7. M. Birkhoj, C. Nellemann, K. Jarfelt, H. Jacobsen, H. R. Andersen, M. Dalgaard and A. M. Vinggaard, The combined antiandrogenic effects of five commonly used pesticides, *Toxicol. Appl. Pharmacol.*, 2004, **201**(1), 10–20.
8. T. Kruger, M. Long and E. C. Bonefeld-Jorgensen, Plastic components affect the activation of the aryl hydrocarbon and the androgen receptor, *Toxicology*, 2008, **246**(2-3), 112–123.
9. M. Long, P. Laier, A. M. Vinggaard, H. R. Andersen, J. Lynggaard and E. C. Bonefeld-Jorgensen, Effects of currently used pesticides in the AhR-CALUX assay: comparison between the human TV101L and the rat H4IIE cell line, *Toxicology*, 2003, **194**(1-2), 77–93.
10. J. Payne, M. Scholze and A. Kortenkamp, Mixtures of four organochlorines enhance human breast cancer cell proliferation, *Environ. Health Perspect.*, 2001, **109**(4), 391–397.
11. N. Rajapakse, E. Silva and A. Kortenkamp, Combining xenoestrogens at levels below individual no-observed-effect concentrations dramatically enhances steroid hormone action, *Environ. Health Perspect.*, 2002, **110**(9), 917–921.
12. S. B. Metzdorff, M. Dalgaard, S. Christiansen, M. Axelstad, U. Hass, M. K. Kiersgaard, M. Scholze, A. Kortenkamp and A. M. Vinggaard, Dysgenesis and histological changes of genitals and perturbations of gene expression in male rats after in utero exposure to antiandrogen mixtures, *Toxicol. Sci.*, 2007, **98**(1), 87–98.
13. M. F. Fernandez, A. Rivas, F. Olea-Serrano, I. Cerrillo, J. M. Molina-Molina, P. Araque, J. L. Martinez-Vidal and N. Olea, Assessment of total effective xenoestrogen burden in adipose tissue and identification of chemicals responsible for the combined estrogenic effect, *Anal. Bioanal. Chem.*, 2004, **379**(1), 163–170.
14. J. Ibarluzea Jr, M. F. Fernandez, L. Santa-Marina, M. F. Olea-Serrano, A. M. Rivas, J. J. Aurrekoetxea, J. Exposito, M. Lorenzo, P. Torne, M. Villalobos, V. Pedraza, A. J. Sasco and N. Olea, Breast cancer risk and the combined effect of environmental estrogens, *Cancer Causes Control*, 2004, **15**(6), 591–600.
15. A. Rivas, M. F. Fernandez, I. Cerrillo, J. Ibarluzea, M. F. Olea-Serrano, V. Pedraza and N. Olea, Human exposure to endocrine disrupters: standardisation of a marker of estrogenic exposure in adipose tissue, *Apmis*, 2001, **109**(3), 185–197.
16. E. C. Bonefeld-Jorgensen, P. S. Hjelmborg, T. S. Reinert, B. S. Andersen, V. Lesovoy, C. H. Lindh, L. Hagmar, A. Giwercman, M. Erlandsen, G. C. Manicardi, M. Spano, G. Toft and J. P. Bonde, Xenoestrogenic activity in blood of European and Inuit populations, *Environ. Health*, 2006, **5**(1), 12.
17. T. Krüger, P. S. Hjelmborg, B. A. G. Jönsson, L. Hagmar, A. Giwercman, G.-C. Manicardi, D. Bizzaro, M. Spanò, A. Rignell-Hydbom, H. S.

Pedersen, G. Toft, J. P. Bonde and E. C. Bonefeld-Jorgensen, Xeno-androgenic activity in serum differs across European and Inuit populations, *Environ. Health Perspect.*, 2007, **115**(suppl. 1), 21–27.
18. M. Long, B. S. Andersen, C. H. Lindh, L. Hagmar, A. Giwercman, G. C. Manicardi, D. Bizzaro, M. Spano, G. Toft, H. S. Pedersen, V. Zvyezday, J. P. Bonde and E. C. Bonefeld-Jorgensen, Dioxin-like activities in serum across European and Inuit populations, *Environ. Health*, 2006, **5**, 14.
19. P. S. Hjelmborg, M. Ghisari and E. C. Bonefeld-Jorgensen, SPE-HPLC purification of endocrine-disrupting compounds from human serum for assessment of xenoestrogenic activity, *Anal. Bioanal. Chem.*, 2006, **385**(5), 875–887.
20. T. H. Rasmussen, F. Nielsen, H. R. Andersen, J. B. Nielsen, P. Weihe and P. Grandjean, Assessment of xenoestrogenic exposure by a biomarker approach: application of the E-Screen bioassay to determine estrogenic response of serum extracts, *Environ. Health*, 2003, **2**(1), 12.
21. E. C. Bonefeld-Jorgensen, H. T. Grunfeld and I. M. Gjermandsen, Effect of pesticides on estrogen receptor transactivation in vitro: a comparison of stable transfected MVLN and transient transfected MCF-7 cells, *Mol. Cell Endocrinol.*, 2005, **244**(1-2), 20–30.
22. E. C. Bonefeld-Jorgensen, M. Long, M. V. Hofmeister and A. M. Vinggaard, Endocrine-disrupting potential of bisphenol A, bisphenol A dimethacrylate, 4-n-nonylphenol, and 4-n-octylphenol *in vitro*: new data and a brief review, *Environ. Health Perspect.*, 2007, **115**(Suppl 1), 69–76.
23. T. Kruger, M. Ghisari, P. S. Hjelmborg, B. Deutch and E. C. Bonefeld-Jorgensen, Xenohormone transactivities are inversely associated to serum POPs in Inuit, *Environ. Health*, 2008, **7**, 38.
24. M. Long, B. Deutch and E. C. Bonefeld-Jørgensen, AhR transcriptional activity in serum of Inuits across Greenlandic districts, *Environ. Health*, 2007, **6**, 1–17.
25. H. H. Samuels and L. E. Shapiro, Thyroid hormone stimulates de novo growth hormone synthesis in cultured GH1 cells: evidence for the accumulation of a rate limiting RNA species in the induction process, *Proc. Natl. Acad. Sci. USA*, 1976, **73**(10), 3369–3373.
26. M. Schriks, C. M. Vrabie, A. C. Gutleb, E. J. Faassen, I. M. C. M. Rietjens and A. J. Murk, T-screen to quantify functional potentiating, antagonistic and thyroid hormone-like activities of poly halogenated aromatic hydrocarbons (PHAHs), *Toxicol. in vitro*, 2006, **20**(4), 490–498.
27. M. Ghisari and E. C. Bonefeld-Jorgensen, Impact of environmental chemicals on the thyroid hormone function in pituitary rat GH3 cells, *Mol. Cell Endocrinol.*, 2005, **244**(1-2), 31–41.
28. H. J. Drenth, C. A. Bouwman, W. Seinen and M. Van den Berg, Effects of some persistent halogenated environmental contaminants on aromatase (CYP19) activity in the human choriocarcinoma cell line JEG-3, *Toxicol. Appl. Pharmacol.*, 1998, **148**(1), 50–55.
29. J. Payne, N. Rajapakse, M. Wilkins and A. Kortenkamp, Prediction and assessment of the effects of mixtures of four xenoestrogens, *Environ. Health Perspect.*, 2000, **108**(10), 983–987.

30. A. Kortenkamp, Ten years of mixing cocktails: a review of combination effects of endocrine-disrupting chemicals, *Environ. Health Perspect.*, 2007, **115**(Suppl 1), 98–105.
31. M. Ghisari and E. C. Bonefeld-Jorgensen, Effects of plasticizers and their mixtures on estrogen receptor and thyroid hormone functions, *Toxicol. Lett.*, 2009, **189**(1), 67–77.
32. J. Butler Walker, L. Seddon, E. McMullen, J. Houseman, K. Tofflemire, A. Corriveau, J. P. Weber, C. Mills, S. Smith and J. Van Oostdam, Organochlorine levels in maternal and umbilical cord blood plasma in Arctic Canada, *Sci. Total Environ.*, 2003, **302**(1-3), 27–52.
33. L. L. Needham, D. B. Barr, S. P. Caudill, J. L. Pirkle, W. E. Turner, J. Osterloh, R. L. Jones and E. J. Sampson, Concentrations of environmental chemicals associated with neurodevelopmental effects in U.S. population, *Neurotoxicology*, 2005, **26**(4), 531–545.
34. A. W. Glynn, A. Wolk, M. Aune, S. Atuma, S. Zettermark, M. Maehle-Schmid, P. O. Darnerud, W. Becker, B. Vessby and H. O. Adami, Serum concentrations of organochlorines in men: a search for markers of exposure. *Sci. Total Environ.*, 2000, **263**(1-3), 197–208.
35. T. J. Schrader and G. M. Cooke, Effects of Aroclors and individual PCB congeners on activation of the human androgen receptor in vitro, *Reprod. Toxicol.*, 2003, **17**(1), 15–23.
36. A. Brouwer, D. C. Morse, M. C. Lans, A. G. Schuur, A. J. Murk, E. Klasson-Wehler, A. Bergman and T. J. Visser, Interactions of persistent environmental organohalogens with the thyroid hormone system: mechanisms and possible consequences for animal and human health, *Toxicol. Ind. Health*, 1998, **14**(1-2), 59–84.
37. J. P. Giesy and K. Kannan, Dioxin-like and non-dioxin-like toxic effects of polychlorinated biphenyls (PCBs): implications for risk assessment, *Crit. Rev. Toxicol.*, 1998, **28**(6), 511–569.
38. J. Suh, J. S. Kang, K. H. Yang and N. E. Kaminski, Antagonism of aryl hydrocarbon receptor-dependent induction of CYP1A1 and inhibition of IgM expression by di-ortho-substituted polychlorinated biphenyls, *Toxicol. Appl. Pharmacol.*, 2003, **187**(1), 11–21.
39. A. O. Cheek, K. Kow, J. Chen and J. A. McLachlan, Potential mechanisms of thyroid disruption in humans: interaction of organochlorine compounds with thyroid receptor, transthyretin, and thyroid-binding globulin, *Environ. Health Perspect.*, 1999, **107**(4), 273–278.
40. H. R. Andersen, A. M. Andersson, S. F. Arnold, H. Autrup, M. Barfoed, N. A. Beresford, P. Bjerregaard, L. B. Christiansen, B. Gissel, R. Hummel, E. B. Jorgensen, B. Korsgaard, R. Le Guevel, H. Leffers, J. McLachlan, A. Moller, J. B. Nielsen, N. Olea, A. Oles-Karasko, F. Pakdel, K. L. Pedersen, P. Perez, N. E. Skakkeboek, C. Sonnenschein and A. M. Soto, Comparison of short-term estrogenicity tests for identification of hormone-disrupting chemicals, *Environ. Health Perspect.*, 1999, **107**(Suppl. 1), 89–108.
41. C. Schmutzler, I. Gotthardt, P. J. Hofmann, B. Radovic, G. Kovacs, L. Stemmler, I. Nobis, A. Bacinski, B. Mentrup, P. Ambrugger, A. Gruters,

L. K. Malendowicz, J. Christoffel, H. Jarry, D. Seidlova-Wuttke, W. Wuttke and J. Kohrle, Endocrine disruptors and the thyroid gland – a combined in vitro and in vivo analysis of potential new biomarkers, *Environ. Health Perspect.*, 2007, **115**(Suppl. 1), 77–83.
42. R. Waring. *Dysregulation of Endogenous Steroid Metabolism Potentially Alters Neuronal and Reproductive System Development: Effects of Environmental Plasticisers*, 2006 [cited], available from: http://ec.europa.eu/research/quality-of-life/ka4/pdf/report_endomet_en.pdf.
43. B. Deutch, H. S. Pedersen, G. Asmund and J. C. Hansen, Contaminants, diet, plasma fatty acids and smoking in Greenland 1999-2005, *Sci. Total Environ.*, 2007, **372**(2-3), 486–496.
44. E. C. Bonefeld-Jorgensen, Biomonitoring in Greenland: human biomarkers of exposure and effects – a short review, *Rural Remote Health*, **10**(2), 1362.
45. M. Pliskova, J. Vondracek, R. F. Canton, J. Nera, A. Kocan, J. Petrik, T. Trnovec, T. Sanderson, M. van den Berg and M. Machala, Impact of polychlorinated biphenyls contamination on estrogenic activity in human male serum, *Environ. Health Perspect.*, 2005, **113**(10), 1277–1284.
46. H. R. Andersen, F. Nielsen, J. B. Nielsen, M. B. Kjaerstad, J. Baelum and P. Grandjean, Xeno-oestrogenic activity in serum as marker of occupational pesticide exposure, *Occup. Environ. Med.*, 2007, **64**(10), 708–714.
47. J. P. Bonde, G. Toft, L. Rylander, A. Rignell-Hydbom, A. Giwercman, M. Spano, G. C. Manicardi, D. Bizzaro, J. K. Ludwicki, V. Zvyezday, E. C. Bonefeld-Jorgensen, H. S. Pedersen, B. A. Jonsson and A. M. Thulstrup, Fertility and markers of male reproductive function in Inuit and European populations spanning large contrasts in blood levels of persistent organochlorines, *Environ. Health Perspect.*, 2008, **116**(3), 269–277.
48. A. Stronati, G. C. Manicardi, M. Cecati, M. Bordicchia, L. Ferrante, M. Spano, G. Toft, J. P. Bonde, B. A. Jonsson, A. Rignell-Hydbom, L. Rylander, A. Giwercman, H. S. Pedersen, E. C. Bonefeld-Jorgensen, J. K. Ludwicki, V. Lesovoy, D. Sakkas and D. Bizzaro, Relationships between sperm DNA fragmentation, sperm apoptotic markers and serum levels of CB-153 and p,p'-DDE in European and Inuit populations, *Reproduction*, 2006, **132**(6), 949–958.
49. T. Kruger, M. Spano, M. Long, P. Eleuteri, M. Rescia, P. S. Hjelmborg, G. C. Manicardi, D. Bizzaro, A. Giwercman, G. Toft, J. P. Bonde and E. C. Bonefeld-Jorgensen, Xenobiotic activity in serum and sperm chromatin integrity in European and inuit populations, *Mol. Reprod. Dev.*, 2008, **75**(4), 669–680.
50. M. Long, A. Stronati, D. Bizzaro, T. Kruger, G. C. Manicardi, P. S. Hjelmborg, M. Spano, A. Giwercman, G. Toft, J. P. Bonde and E. C. Bonefeld-Jorgensen, Relation between serum xenobiotic-induced receptor activities and sperm DNA damage and sperm apoptotic markers in European and Inuit populations, *Reproduction*, 2007, **133**(2), 517–530.

CHAPTER 11

Biomarkers of Exposure: Oxidative Stress to DNA and Lipids – Relation to Air Pollution

STEFFEN LOFT* AND PETER MØLLER

University of Copenhagen, Department of Public Health, Section of Environmental Health, Øster Farimagsgade 5A, Postbox 2099, DK-1014, Copenhagen K, Denmark

11.1 Introduction: Oxidative Stress and Disease

11.1.1 Oxidative Stress

The term oxidative stress has >80 000 entries in PUBMED and it has traditionally been defined as an imbalance between the level of oxidants and the antioxidant system in the cell.[1] Reactive oxygen species (ROS) are generated continuously as byproducts of the normal cellular energy production, by daily activities such as exhaustive exercise or in conditions such as obesity, stress, ischaemia/reperfusion, infection and inflammation, which also generate reactive nitrogen species (RNS) (Figure 11.1). Multiple environmental exposures, such as air pollution, radiation, many metals and multiple foreign chemicals, through their metabolism, as well as lifestyle factors including smoking, alcohol consumption and a poor diet, are important external sources of ROS.[2] The antioxidant system defence against ROS encompasses antioxidant enzymes, including superoxide dismutases (SOD), catalase and gluthathione peroxidases (GPx), as well as radical scavengers, such as glutathione (GSH), urate, vitamin C,

Issues in Toxicology No. 10
Biomarkers and Human Biomonitoring, Volume 2: Selected Biomarkers of Current Interest
Edited by Lisbeth E. Knudsen and Domenico Franco Merlo

Published by the Royal Society of Chemistry, www.rsc.org

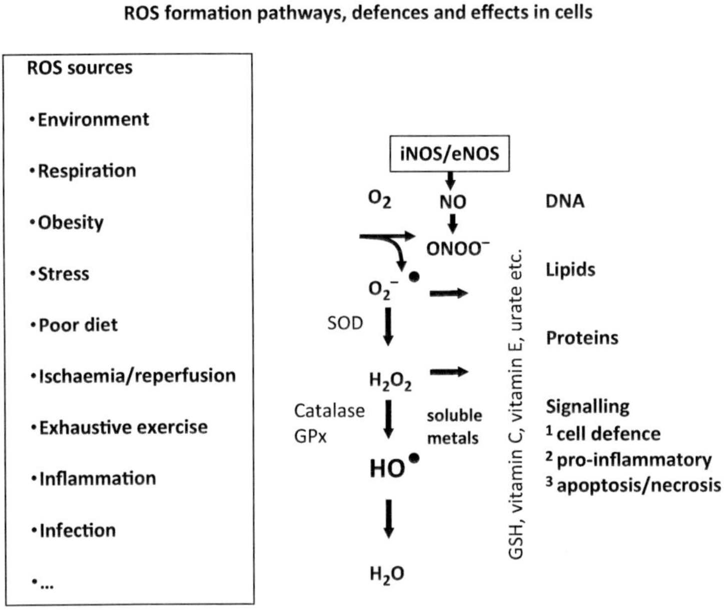

Figure 11.1 Sources of reactive oxygen species (ROS), defence mechanisms and effects in cells. eNOS, endothelial nitric oxide synthase; GPx, glutathione peroxidase; GSH, glutathione; iNOS, inducible nitric oxide synthase; SOD, superoxide dismutase.

vitamin E, *etc.*, many of which are micronutrients or nutrient dependent.[3] Elevated levels of oxidative stress are associated with damage to DNA, protein and lipids as well as activation and perturbation of multiple signalling pathways that inflict upon the normal physiological function of the cell. These pathways have been roughly divided into three levels related to the level of oxidative stress: cellular defence activation, inflammatory activation, and activation of cell death pathways.[4]

11.1.2 Involvement of Oxidative Stress in Disease Pathogenesis

Oxidative stress appears to be involved in the pathogenesis, complications or both of a surprisingly wide variety of diseases, including almost all major public health-related conditions, with a few central examples briefly described herein.[5–7] Thus, oxidative stress leads to mutagenic DNA damage, microsatellite instability and promotion relevant for many cancers. Oxidative stress causes oxidation of lipids and proteins, inflammation and endothelial dysfunction which are all central in the atherosclerotic and thrombogenic processes of vascular diseases. Even the ischaemia–reperfusion injury to the tissues in myocardial infarction and stroke is partly related to oxidative stress. The killing of pancreatic beta cells in diabetes is mediated by oxidative and nitrosative stress, whereas many of the diabetic complications are mediated by oxidative stress, as demonstrated by related damage to DNA, RNA, proteins and lipids, in particular in patients with poor glycaemic control. Oxidative changes are found in the brain tissue of the major neurodegenerative diseases such as Alzheimer and Parkinson diseases, and this is thought to play a pathogenic role. Many of the processes of ageing involve oxidative stress; oxidative damage to mitochondrial DNA leads to excess mitochondrial generation of ROS in a vicious circle, and many mitochondrial (dys)functions are related to ageing. Similarly, oxidative stress can cause telomere shortening and apoptosis. Lastly, reproduction is highly susceptible to oxidative stress; oxidation of sperm DNA reduces fertility and can result in germ-line mutations, whereas fetal exposure to oxidative stress is likely to be a wide ranging mechanism of minor and major congenital malfunctions.

However, it should also be recognized that many relevant cell processes, including cell growth, migration and differentiation, are regulated by fine-tuned redox balances and that a central part of our innate immune system relies on oxidative stress agents to kill invading microorganisms.

11.2 Biomarkers of Oxidative Stress to DNA and Lipids

11.2.1 Matrix of Biomarkers of Oxidative Stress

Oxidation products of lipids and DNA can be considered as biomarkers of the biologically effective dose of oxidative stress (Figure 11.2).[8] The DNA in any available cell can be assessed for oxidative damage, but surrogate cells such as peripheral blood mononuclear cells (PBMC) are usually studied, whereas water soluble repair products can in principle be measured in any biological fluid. Lipid peroxidation products can be assessed in any biological fluid, in exhaled breath or in tissues/cells. The specificity and validity of the biomarker reflects both methodological properties and the biomarker's ability to detect events in biological or toxicological mechanisms. Therefore, the specificity and validity refer to the ability of these biomarkers to assess the magnitude of oxidative stress adequately in the relevant context.

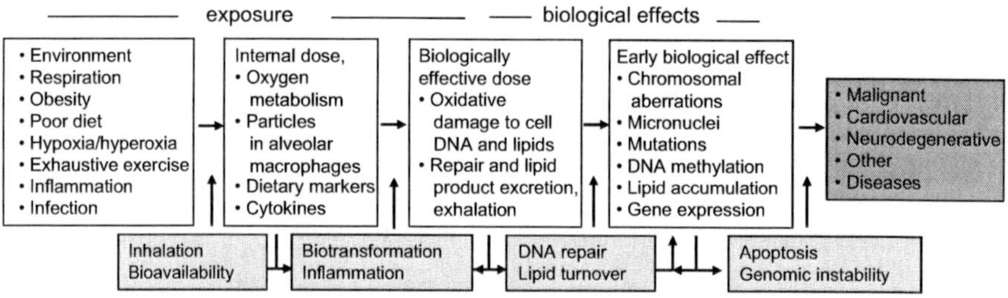

Figure 11.2 Biomarkers of exposure and effects of oxidative stress.

11.2.2 Biomarkers of DNA Damage from Oxidative Stress

ROS generate a large number of oxidative modifications in DNA, including strand breaks (SB) and base oxidations.[9–12] Among such DNA damage products, 8-oxo-7,8-dihydro-2′-deoxyguanosine (8-oxodG) is probably the most studied oxidation product, owing to its relative ease of measurement and premutagenic potential.[13,14] Biomarkers of oxidative damage to DNA encompass methods that measure breaks in the DNA strand and base oxidation products.

Although exposure to oxidizing agents generates SB, the assessment of SB in tissues and PBMC is usually carried out by methods that measure overall genotoxicity rather than SB. The single cell electrophoresis (comet) assay has become a very popular and sensitive assay for the detection of SB in PBMC of humans exposed to different occupational and environmental agents, but it is not a specific endpoint of oxidatively damaged DNA.[15] The comet assay provides a more reliable measurement of oxidative stress by quantitative assessment of the excess SB induced by enzymatic digestion of DNA with endonuclease III (ENDOIII) or formamidopyrimidine DNA glycosylase (FPG) from *Escherichia coli* that mainly detects ring-opened purine and pyrimidine lesions, respectively. In theory, even higher specificity of the detected lesion should be achieved if human oxoguanine glycosylase (hOGG1) is used for DNA incision, although this has not been validated systematically. A European multicentre validation exercise has recently demonstrated that variation in the comet assay scoring can be reduced by using calibration curves based on ionizing radiation, which also allows translation of any score into lesions per cell or, for example, base pair.[16,17] The comet assay shows limited problems related to spurious oxidation with proper sample handling and storage of cells in medium.[18,19]

The DNA base oxidation products in cells, tissues and urine can be measured by chromatographic methods with electrochemical or mass spectrometry based detection. The latter allows highly specific determination of multiple lesions, particularly by the use of heavy isotope labelled internal standards.[11,20,21] In cells and tissues the chromatographic techniques require extraction of DNA before hydrolysis, usually by enzymatic digestion to nucleoside level, whereas gas chromatography requires both acid hydrolysis and derivatization. The major problem found in measuring oxidized bases in DNA is the occurrence of spurious oxidation during these sample preparation steps.[18] DNA extraction is a critical issue and derivatization is particularly problematic, as described by the European Standards Committee on Oxidative DNA Damage (ESCODD), which also concluded that the true levels of 8-oxodG in cellular DNA are between 0.5 and 5 lesions per 10^6 dG.[18,19]

The urinary excretion of products of damaged nucleotides from cellular pools or DNA may also be important. In a steady state, the urinary excretion of 8-oxodG and similar products should, in principle, reflect the average rate of oxidative damage in the whole body, whereas the level of lesions in DNA from target cells or surrogate cells, such as PBMC, should reflect the balance between damage and repair only in these cell types.[22] Of the many oxidative damage products, 8-oxodG is also the most studied in urine, with chromatographic assays and specific detection based on electrochemical or advanced tandem mass spectrometry, whereas enzyme-linked immunosorbent assay (ELISA) is widely used because of the simple requirement for equipment.[22] The methods have recently been compared systematically, with the chromatographic assays showing consistency, whereas ELISA-based assays showed more variation and systematically higher values.[23]

11.2.3 Biomarkers of Lipid Peroxidation

ROS can attack any lipid in, for example, membranes or plasma, generating a variety of relatively stable end-products, such as malondialdehyde (MDA), 4-hydroxynonenal (HNE), acrolein and a range of isoprostanes.[24] There is a range of biomarkers of lipid peroxidation products. Some of the biomarkers have been severely criticized and should be avoided for *in vivo* detection of lipid peroxidation.[3] These include the simple thiobarbituric acid reactive substances (TBARS) assay that should be dismissed because most TBARS are not related to lipid peroxidation, whereas improved methods using high performance liquid chromatography (HPLC) purification steps are more reliable assays.[25] The specificity of *in vivo* measurements has also been questioned for the conjugated diene assay and the simple assays for determination of lipid hydroperoxides (LH), including the ferrous oxidation–xylenol orange (FOX) assay and similar assays that are based on measurements of absorbance of oxidation products, because they are non-specific measures of lipid peroxidation.[26] The isoprostanes are probably the best available biomarker of lipid peroxidation when measured by mass spectrometry with stable isotope dilution. However, detection of isoprostanes by immunoassays is the most widely used method because such assays are technically simple and cheap.[25,27] An alternative to the measurement of lipid peroxidation products in body fluids is the assessment of hydrocarbons (pentane and ethane) in exhaled air, although the validity and use of this biomarker is hampered because the hydrocarbon gases are minor end products of lipid peroxidation, there is no consensus on the background level, and they are difficult to measure.[25,26,28]

11.2.4 Validation Status of Biomarkers of Oxidative Stress and Potential Bias

The validation status of biomarkers implies a broadly accepted notion among researchers that the method provides reliable measurements that are reproducible and have a predictive value in terms of the health effect. Although a biomarker may be specific, there may not be consensus about normal levels in healthy humans. Currently, it is commonly accepted that the true level of oxidized guanines in DNA is in the range of 0.3–4.2 lesions/10^6 dG and there exist reference values for comet assay endpoints,[29–32] whereas it remains challenging to compare results of urinary excretion of 8-oxodG because the data are reported in different units such as the concentration and total excretion over a period of time. Typical levels of plasma or serum LH and isoprostanes are known.[27,33]

Biomarkers are used as intermediate endpoints that measure events in the mechanism of the disease from exposure to actual outcome. Therefore, the predictive value is an important feature of the validity of biomarkers. Oxidative stress is commonly regarded to be associated with various diseases such as cancer, coronary artery disease and diabetes, but the biomarkers of oxidative stress may be elevated in patients as a consequence of the disease.[34] The predictive value of biomarkers should be evaluated in prospective studies that may involve a biobank-based type of cohort design.[35] Using this type of approach, it has been shown that urinary excretion of 8-oxodG is a risk marker of lung cancer in non-smokers.[36] A number of studies have found elevated or decreased levels of comet assay endpoints, DNA oxidation or repair capacity in various cancer patients in case–control settings, but such findings are highly susceptible to reverse causality.[31,35,37] The predictive value of lipid peroxidation markers has been investigated in a few longitudinal studies. Patients with stable coronary artery disease and haemodialysis patients with high serum levels of TBARS had increased risk of developing cardiovascular events.[38,39] In addition, the predictive value of plasma lipid peroxidation products (denoted as malondialdehyde measured by a commercially available kit) was associated with increased mortality in elderly institutionalized persons.[40] To the best of our knowledge, the predictive value of isoprostanes has not been investigated in prospective studies, although there is growing belief that it is the most valuable measurement of the currently available biomarkers of lipid peroxidation.[33]

11.3 Air Pollution and Biomarkers of Oxidative Stress

11.3.1 Air Pollution and Oxidative Stress

Exposure to air pollution, especially particulate matter (PM), is probably one of the areas with the most widely used and systematic validation of biomarkers of exposure to oxidative stress in DNA and lipids. This is particularly relevant because exposure to PM from combustion processes contributes substantially to, in particular, cardiovascular and pulmonary ill health, lung cancer and premature mortality globally.[41,42] Moreover, oxidative stress is mechanistically important in cardiovascular disease through lipid oxidation, inflammation, endothelial dysfunction and a wide variety of related mechanisms. Similarly, oxidative stress is central in airway disease and lung cancer.[4,43,44]

PM can cause oxidative stress through multiple mechanisms depicted in Figure 11.3, including: (i) direct generation of ROS by physico-chemical surface properties of particles, (ii) effects of soluble compounds such as transition metals or organic compounds, (iii) effects of particles on mitochondria, NADPH-oxidases or cellular calcium homeostasis, and (iv) activation of inflammatory cells capable of generating ROS and RNS. Oxidative

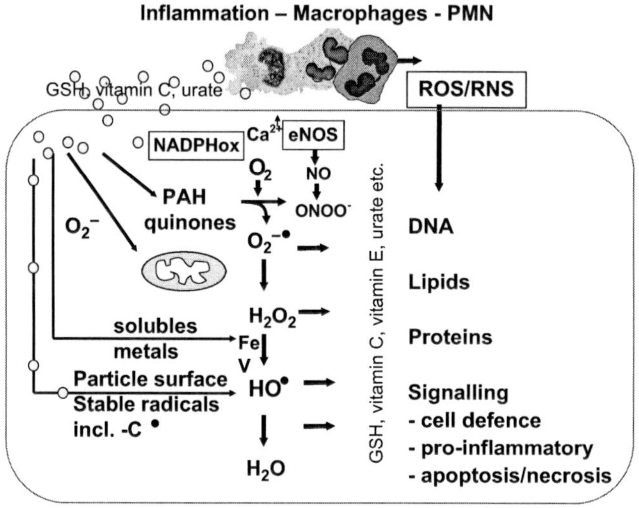

Figure 11.3 Generation of oxidative stress and damage in cells from exposure to particulate matter.

stress is an important trigger of PM-induced inflammation through activation of redox sensitive signalling pathways, including AP-1, MAPkinases and NF-κB activation with subsequent proinflammatory gene expression, although direct activation of MAPkinases through EGRF also occurs.[45] Ultrafine particles of less than 100 nm in aerodynamic diameter are particularly prone to induce oxidative stress through these pathways because of high alveolar deposition, more cellular uptake and large surface area for chemical reactions. PM not only represents highly complex mixtures with large variation in size, chemical composition, shape, surface, reactivity and charge in both time and space due to variable sources, atmospheric chemical reactions and meteorological conditions,[46] there is usually accompanying exposure to other direct and indirect oxidants such as ozone and NO_2 or benzene from which metabolism generates ROS.[8]

There is vast experimental evidence for the role of oxidative stress and damage to DNA and lipids in relation to effects of PM in cancer and vascular diseases.[43,47,48]

11.3.2 Systematic Review of Biomarkers of Oxidative Stress in Relation to Air Pollution

A number of studies of PM exposure in humans have applied biomarkers of oxidative damage to DNA and lipids in the blood compartment or in terms of products excreted in urine or exhaled breath condensate (EBC).[8] The biomarkers of oxidatively damaged DNA include 8-oxodG or the corresponding base, 8-oxo-7,8-dihydroguanine (8-oxoGua) measured in DNA and urine, the exocyclic M1 adduct to guanine (M1dG), a result of adduction of lipid peroxidation products to DNA, and lesions detected as FPG and ENDOIII sensitive sites in DNA. The biomarkers of lipid peroxidation (LPO) products include conjugated dienes (CD), lipid hydroperoxides (MDA; TBARS) and F2-isoprostanes measured in EBC, plasma, serum or urine. An exhaustive set 46 of these studies has recently been assessed systematically by meta-analysis tools in order to evaluate the validity of the use of these biomarkers of biological effective dose in relation to air pollution.[8]

The qualitative assessment indicated that most of the published reports show associations between air pollution exposure and oxidatively damaged DNA, nucleobases and lipids.[8] The majority of the studies have measured the biomarkers in surrogate tissue cells such as white blood cells (WBC) or non-cellular body fluids such as plasma, urine and EBC. The data on biomarkers of the airways mainly encompass measurements of LPO products in EBC, whereas there is only one study on 8-oxodG in nasal cells.[49]

The effect of exposure to air pollution assessed as the standardized mean difference of all available studies for the oxidized DNA and LPO products in the blood was 0.53 (95% confidence interval, CI: 0.29–0.76) and 0.73 (95% CI: 0.18–1.28), respectively. These numbers indicate that the mean level of the biomarker increased by 53% and 73% for DNA and LPO products upon exposure to air pollution on average across the studies. In the urine the effect size for PM exposure was 0.49 (95% CI: 0.01–0.97) and 0.57 (95% CI: 0.22–0.82) as standardized mean differences for the LPO products and oxidatively damaged DNA and nucleobases, respectively. This implies that exposure to PM generates the same effect size in oxidized DNA and lipids, whereas the effect appears larger in the blood than the urine. This pattern correlates with the level of exposure because it must be expected that WBC and plasma circulate through the lungs and intestine where they may come into contact with particles through inhalation and ingestion of PM, respectively. The effect on DNA damage in the airways is currently difficult to assess because there is only one study of oxidized DNA.[49] The effect on LPO products in EBC was 0.64 (95% CI: 0.07–1.21), which is not different from the results obtained in the blood and urine. This could indicate that LPO products in plasma and urine are suitable biomarkers of biological effective dose, reflecting oxidative stress in the airways.

The systematic assessment also included assessment of potential bias from poor exposure assessment or control by design and the use of biomarkers with low specificity.[8] The associations and effect estimates were seen across studies with optimum design, including controlled or personal exposure assessment and biomarkers with low risk of artefacts, to studies with risk of bias due to indirect exposure assessment and/or use of biomarkers prone to artefacts. Still, it should be emphasized that the studies identified were inhomogeneous in design and quality of biomarkers, which weakens the conclusion about specific exposure–effect relationships in regard to particulate air pollution.

11.4 Biomarkers of Oxidative Stress and Exposures through Occupation, Drinking Water and Diet

Many occupational exposures result in oxidative stress with damage to DNA and lipids, as reviewed by others.[50,51] Similarly, health-threatening exposure to arsenic through drinking water occurs in many parts of the world. Part of the mechanism of action of arsenic involves oxidative stress, as recently outlined in a review with a focus on biomarkers.[52] The role of oxidative stress in the health effects of environmental exposure in diet can also be addressed by biomarkers.[2]

It is also important to control environmental exposure studies for the possible protective effects against oxidative damage to DNA and lipids of diets rich in or supplements of antioxidants, which has been addressed in a large number of studies as outlined in several reviews.[53–57] The biomarkers of DNA damage have included oxidized bases in leukocytes or urinary excretion of 8-oxodG.[55] Of 145 intervention studies identified, 45 were of good design quality with control of confounding from other time-dependent factors. Typically, parallel group or cross-over designs in random orders were used. More than one third of the intervention studies showed reduced levels of damage during intervention. This is significantly more than expected by chance, and similar effect size and power in studies with and without report of significant effects suggest that this is not likely to be due to publication bias.[55,56] Although not statistically significant there are trends that suggest that antioxidant foods are more efficient than supplements, especially with respect to urinary excretion of 8-oxodG, and that the effects are more likely to be demonstrated in subjects with oxidative stress.[56] Nevertheless, the real problem may be that very few studies have statistical power to detect effects that are of realistic size, *i.e.* reductions of 10% or less, and the large differences in protocols preclude use of meta-analysis.

Smokers have consistently elevated urinary excretion of 8-oxodG, whereas the levels in lymphocytes appears to be more variable.[35] This could be due to different regulation of the repair mechanisms among smokers and non-smokers, although such studies of repair activity have shown mixed results and there might be interactions with nutritional status and antioxidant intake.[58–60]

11.5 Conclusion

Oxidative stress can be caused by many different environmental exposures and it is involved in the pathogenesis of multiple important diseases, in the causative pathways and/or in the progression or complications. Thus, biomarkers of exposure to relevant biologically effective oxidative stress are important. So far, most focus has been on biomarkers of oxidative damage to DNA and lipids measured in cells, tissues or a range of biological fluids. Exposures such as air pollution have been well validated to cause increases in such biomarkers, which are relevant to outcomes attributed to this exposure, especially cancer and airway and cardiovascular diseases. However, it should also be recognized that biomarkers of oxidative stress to DNA and lipids are not specific for any exposure or for any disease outcome. Moreover, diet, physiological processes and repair mechanisms and kinetics can influence the levels of biomarkers, which needs to be taken into account in the interpretation of data.

Acknowledgements

The work of Steffen Loft and Peter Møller is supported by The Danish Research Councils, The European Commission and Realdania through CISBO.

References

1. H. Sies, Oxidative stress: oxidants and antioxidants, *Exp. Physiol*, 1997, **82**, 291–295.
2. D. Ziech, R. Franco, A. G. Georgakilas, S. Georgakila, V. Malamou-Mitsi, O. Schoneveld, A. Pappa and M. I. Panayiotidis, The role of reactive oxygen species and oxidative stress in environmental carcinogenesis and biomarker development, *Chem. Biol. Interact.*, 2010, **188**, 334–339.
3. B. Halliwell and J. M. Gutteridge, *Free Radicals in Biology and Medicine*, Oxford University Press, Oxford, 1999.
4. N. Li, T. Xia and A. E. Nel, The role of oxidative stress in ambient particulate matter-induced lung diseases and its implications in the toxicity of engineered nanoparticles, *Free Radic. Biol. Med.*, 2008, **44**, 1689–1699.
5. B. Halliwell, Free radicals and antioxidants – quo vadis? *Trends Pharmacol. Sci.*, 2011, **32**, 125–130.
6. R. A. Roberts, D. L. Laskin, C. V. Smith, F. M. Robertson, E. M. Allen, J. A. Doorn and W. Slikker, Nitrative and oxidative stress in toxicology and disease, *Toxicol. Sci.*, 2009, **112**, 4–16.
7. A. Negre-Salvayre, N. Auge, V. Ayala, H. Basaga, J. Boada, R. Brenke, S. Chapple, G. Cohen, J. Feher, T. Grune, G. Lengyel, G. E. Mann, R. Pamplona, G. Poli, M. Portero-Otin, Y. Riahi, R. Salvayre, S. Sasson, J. Serrano, O. Shamni, W. Siems, R. C. Siow, I. Wiswedel, K. Zarkovic and N. Zarkovic, Pathological aspects of lipid peroxidation, *Free Radic. Res.*, 2010, **44**, 1125–1171.
8. P. Moller and S. Loft, Oxidative damage to DNA and lipids as biomarkers of exposure to air pollution, *Environ. Health Perspect.*, 2010, **118**, 1126–1136.
9. J. Cadet, S. Bellon, M. Berger, A. G. Bourdat, T. Douki, V. Duarte, S. Frelon, D. Gasparutto, E. Muller, J. L. Ravanat and S. Sauvaigo, Recent aspects of oxidative DNA damage: guanine lesions, measurement and substrate specificity of DNA repair glycosylases, *Biol. Chem.*, 2002, **383**, 933–943.
10. M. Dizdaroglu, Oxidative damage to DNA in mammalian chromatin, *Mutat. Res.*, 1992, **275**, 331–342.

11. M. Dizdaroglu, P. Jaruga, M. Birincioglu and H. Rodriguez, Free radical-induced damage to DNA: mechanisms and measurement, *Free Radic. Biol. Med.*, 2002, **32**, 1102–1115.
12. S. Bjelland and E. Seeberg, Mutagenicity, toxicity and repair of DNA base damage induced by oxidation, *Mutat. Res.*, 2003, **531**, 37–80.
13. H. Kasai, Analysis of a form of oxidative DNA damage, 8-hydroxy-2′-deoxyguanosine, as a marker of cellular oxidative stress during carcinogenesis, *Mutat. Res.*, 1997, **387**, 147–163.
14. M. S. Cooke, S. Loft, R. Olinski, M. D. Evans, K. Bialkowski, J. R. Wagner, P. C. Dedon, P. Moller, M. M. Greenberg and J. Cadet, Recommendations for standardized description of and nomenclature concerning oxidatively damaged nucleobases in DNA, *Chem. Res. Toxicol.*, 2010, **23**, 705–707.
15. P. Møller, L. E. Knudsen, S. Loft and H. Wallin, The comet assay as a rapid test in biomonitoring occupational exposure to DNA-damaging agents and effect of confounding factors, *Cancer Epidemiol. Biomarkers Prev.*, 2000, **9**, 1005–1015.
16. L. Forchhammer, E. V. Bräuner, J. K. Folkmann, P. H. Danielsen, C. Nielsen, A. Jensen, S. Loft, G. Friis, and P. Moller, Variation in assessment of oxidatively damaged DNA in mononuclear blood cells by the comet assay with visual scoring, *Mutagenesis*, 2008, **23**, 223–231.
17. C. Johansson, P. Moller, L. Forchhammer, S. Loft, R. W. Godschalk, S. A. Langie, S. Lumeij, G. D. Jones, R. W. Kwok, A. Azqueta, D. H. Phillips, O. Sozeri, M. N. Routledge, A. J. Charlton, P. Riso, M. Porrini, A. Allione, G. Matullo, J. Palus, M. Stepnik, A. R. Collins and L. Moller, An ECVAG trial on assessment of oxidative damage to DNA measured by the comet assay, *Mutagenesis*, 2010, **25**, 125–132.
18. European Standards Committee on Oxidative DNA Damage (ESCODD), Measurement of DNA oxidation in human cells by chromatographic and enzymic methods, *Free Radic. Biol. Med.*, 2003, **34**, 1089–1099.
19. C. M. Gedik and A. Collins, Establishing the background level of base oxidation in human lymphocyte DNA: results of an interlaboratory validation study, *FASEB J.*, 2005, **19**, 82–84.
20. J. Cadet, T. Douki, S. Frelon, S. Sauvaigo, J. P. Pouget and J. L. Ravanat, Assessment of oxidative base damage to isolated and cellular DNA by HPLC-MS/MS measurement, *Free Radic. Biol. Med.*, 2002, **33**, 441–449.
21. A. Weimann, D. Belling and H. E. Poulsen, Quantification of 8-oxo-guanine and guanine as the nucleobase, nucleoside and deoxynucleoside forms in human urine by high-performance liquid chromatography-electrospray tandem mass spectrometry, *Nucleic Acids Res.*, 2002, **30**, E7.
22. M. S. Cooke, R. Olinski and S. Loft, Measurement and meaning of oxidatively modified DNA lesions in urine, *Cancer Epidemiol. Biomarkers Prev.*, 2008, **17**, 3–14.
23. M. D. Evans, R. Olinski, S. Loft and M. S. Cooke, Toward consensus in the analysis of urinary 8-oxo-7,8-dihydro-2′-deoxyguanosine as a non-invasive biomarker of oxidative stress, *FASEB J.*, 2010, **24**, 1249–1260.

24. P. Moller, L. Risom, C. Lundby, L. Mikkelsen and S. Loft, Hypoxia and oxidation levels of DNA and lipids in humans and animal experimental models, *IUBMB Life*, 2008, **60**, 707–723.
25. B. Halliwell and M. Whiteman, Measuring reactive species and oxidative damage *in vivo* and in cell culture: how should you do it and what do the results mean? *Br. J. Pharmacol.*, 2004, **142**, 231–255.
26. H. R. Griffiths, L. Moller, G. Bartosz, A. Bast, C. Bertoni-Freddari, A. Collins, M. Cooke, S. Coolen, G. Haenen, A. M. Hoberg, S. Loft, J. Lunec, R. Olinski, J. Parry, A. Pompella, H. Poulsen, H. Verhagen and S. B. Astley, Biomarkers, *Mol. Aspects Med.*, 2002, **23**, 101–208.
27. G. L. Milne, E. S. Musiek and J. D. Morrow, F2-isoprostanes as markers of oxidative stress *in vivo*: an overview, *Biomarkers*, 2005, **10**(Suppl 1), S10–S23.
28. E. A. Meagher and G. A. Fitzgerald, Indices of lipid peroxidation *in vivo*: strengths and limitations, *Free Radic. Biol. Med.*, 2000, **28**, 1745–1750.
29. A. R. Collins, The comet assay for DNA damage and repair: principles, applications, and limitations, *Mol. Biotechnol.*, 2004, **26**, 249–261.
30. A. R. Collins, J. Cadet, L. Moller, H. E. Poulsen and J. Vina, Are we sure we know how to measure 8-oxo-7,8-dihydroguanine in DNA from human cells? *Arch. Biochem. Biophys.*, 2004, **423**, 57–65.
31. P. Moller, The alkaline comet assay: towards validation in biomonitoring of DNA damaging exposures, *Basic Clin. Pharmacol. Toxicol.*, 2006, **98**, 336–345.
32. P. Moller, Assessment of reference values for DNA damage detected by the comet assay in human blood cell DNA, *Mutat. Res.*, 2006, **612**, 84–104.
33. B. Halliwell, Lipid peroxidation, antioxidants and cardiovascular disease: how should we move forward? *Cardiovasc. Res.*, 2000, **47**, 410–418.
34. R. Olinski, D. Gackowski, R. Rozalski, M. Foksinski and K. Bialkowski, Oxidative DNA damage in cancer patients: a cause or a consequence of the disease development? *Mutat. Res.*, 2003, **531**, 177–190.
35. S. Loft and P. Moller, Oxidative DNA damage and human cancer: need for cohort studies, *Antioxid. Redox. Signal.*, 2006, **8**, 1021–1031.
36. S. Loft, P. Svoboda, H. Kasai, A. Tjonneland, U. Vogel, P. Moller, K. Overvad and O. Raaschou-Nielsen, Prospective study of 8-oxo-7,8-dihydro-2′-deoxyguanosine excretion and the risk of lung cancer, *Carcinogenesis*, 2006, **27**, 1245–1250.
37. P. Moller, Assessment of reference values for DNA damage detected by the comet assay in human blood cell DNA, *Mutat. Res.*, 2006, **612**, 84–104.
38. M. Boaz, Z. Matas, A. Biro, Z. Katzir, M. Green, M. Fainaru and S. Smetana, Comparison of hemostatic factors and serum malondialdehyde as predictive factors for cardiovascular disease in hemodialysis patients, *Am. J. Kidney Dis.*, 1999, **34**, 438–444.
39. M. F. Walter, R. F. Jacob, B. Jeffers, M. M. Ghadanfar, G. M. Preston, J. Buch and R. P. Mason, Serum levels of thiobarbituric acid reactive substances predict cardiovascular events in patients with stable coronary

artery disease: a longitudinal analysis of the PREVENT study, *J. Am. Coll. Cardiol.*, 2004, **44**, 1996–2002.
40. J. M. Huerta, S. Gonzalez, S. Fernandez, A. M. Patterson and C. Lasheras, Lipid peroxidation, antioxidant status and survival in institutionalised elderly: a five-year longitudinal study, *Free Radic. Res.*, 2006, **40**, 571–578.
41. R. D. Brook, S. Rajagopalan, C. A. Pope, III, J. R. Brook, A. Bhatnagar, A. V. Diez-Roux, F. Holguin, Y. Hong, R. V. Luepker, M. A. Mittleman, A. Peters, D. Siscovick, S. C. Smith, Jr., L. Whitsel and J. D. Kaufman, Particulate matter air pollution and cardiovascular disease: An update to the scientific statement from the American Heart Association, *Circulation*, 2010, **121**, 2331–2378.
42. S. S. Salvi and P. J. Barnes, Chronic obstructive pulmonary disease in non-smokers, *Lancet*, 2009, **374**, 733–743.
43. P. Moller, N. R. Jacobsen, J. K. Folkmann, P. H. Danielsen, L. Mikkelsen, J. G. Hemmingsen, L. K. Vesterdal, L. Forchhammer, H. Wallin and S. Loft, Role of oxidative damage in toxicity of particulates, *Free Radic. Res.*, 2010, **44**, 1–46.
44. P. Moller, J. K. Folkmann, L. Forchhammer, E. V. Bräuner, P. H. Danielsen, L. Risom and S. Loft, Air pollution, oxidative damage to DNA, and carcinogenesis, *Cancer Lett.*, 2008, **266**, 84–97.
45. R. Duffin, N. L. Mills and K. Donaldson, Nanoparticles – a thoracic toxicology perspective, *Yonsei Med. J*, 2007, **48**, 561–572.
46. C. Sioutas, R. J. Delfino and M. Singh, Exposure assessment for atmospheric ultrafine particles (UFPs) and implications in epidemiologic research, *Environ. Health Perspect.*, 2005, **113**, 947–955.
47. A. M. Knaapen, P. J. Borm, C. Albrecht and R. P. Schins, Inhaled particles and lung cancer. Part A: Mechanisms, *Int. J. Cancer*, 2004, **109**, 799–809.
48. L. Risom, P. Møller and S. Loft, Oxidative stress-induced DNA damage by particulate air pollution, *Mutat. Res.*, 2005, **592**, 119–137.
49. L. Calderon-Garciduenas, L. Wen-Wang, Y. J. Zhang, A. Rodriguez-Alcaraz, N. Osnaya, A. Villarreal-Calderon and R. M. Santella, 8-hydroxy-2′-deoxyguanosine, a major mutagenic oxidative DNA lesion, and DNA strand breaks in nasal respiratory epithelium of children exposed to urban pollution, *Environ. Health Perspect.*, 1999, **107**, 469–474.
50. A. Pilger and H. W. Rudiger, 8-Hydroxy-2′-deoxyguanosine as a marker of oxidative DNA damage related to occupational and environmental exposures, *Int. Arch. Occup. Environ. Health*, 2006, **80**, 1–15.
51. M. Corradi, P. Gergelova and A. Mutti, Use of exhaled breath condensate to investigate occupational lung diseases, *Curr. Opin. Allergy Clin. Immunol.*, 2010, **10**, 93–98.
52. A. De Vizcaya-Ruiz, O. Barbier, R. Ruiz-Ramos and M. E. Cebrian, Biomarkers of oxidative stress and damage in human populations exposed to arsenic, *Mutat. Res.*, 2009, **674**, 85–92.
53. P. Møller and S. Loft, Oxidative DNA damage in human white blood cells in dietary antioxidant intervention studies, *Am. J. Clin. Nutr.*, 2002, **76**, 303–310.

54. P. Moller and S. Loft, Interventions with antioxidants and nutrients in relation to oxidative DNA damage and repair, *Mutat. Res.*, 2004, **551**, 79–89.
55. P. Moller and S. Loft, Dietary antioxidants and beneficial effect on oxidatively damaged DNA, *Free Radic. Biol. Med.*, 2006, **41**, 388–415.
56. S. Loft, P. Moller, M. S. Cooke, R. Rozalski and R. Olinski, Antioxidant vitamins and cancer risk: is oxidative damage to DNA a relevant biomarker? *Eur. J. Nutr.*, 2008, **47**(Suppl. 2), 19–28.
57. L. O. Dragsted, Biomarkers of exposure to vitamins A, C, and E and their relation to lipid and protein oxidation markers, *Eur. J. Nutr.*, 2008, **47**(Suppl. 2), 3–18.
58. S. Asami, T. Hirano, R. Yamaguchi, Y. Tomioka, H. Itoh and H. Kasai, Increase of a type of oxidative DNA damage, 8-hydroxyguanine, and its repair activity in human leukocytes by cigarette smoking, *Cancer Res.*, 1996, **56**, 2546–2549.
59. D. Gackowski, E. Speina, M. Zielinska, J. Kowalewski, R. Rozalski, A. Siomek, T. Paciorek, B. Tudek and R. Olinski, Products of oxidative DNA damage and repair as possible biomarkers of susceptibility to lung cancer, *Cancer Res.*, 2003, **63**, 4899–4902.
60. S. Guarnieri, S. Loft, P. Riso, M. Porrini, L. Risom, H. E. Poulsen, L. O. Dragsted and P. Moller, DNA repair phenotype and dietary antioxidant supplementation, *Br. J Nutr.*, 2008, **99**, 1018–1024.

CHAPTER 12
Male and Female Germ Cell Biomarkers

DIANA ANDERSON,*[,a] JULIAN LAUBENTHAL[a] AND ADOLF BAUMGARTNER[a,b]

[a] University of Bradford, Division of Medical Sciences, Richmond Road, Bradford, BD7 1DP, United Kingdom; [b] Department of Paediatric Cardiology, Heart Centre, University of Leipzig, Struempellstr. 39, Leipzig 04289, Germany

12.1 General Introduction

Biomarkers in germ cells have the same function as in somatic cells; that is, to give an indication of immediate changes at the genomic, transcriptional and translational level in the cell itself and the potential transmission of some of these alterations to the offspring.

There are various types of genetic change which can be detected in germ cells. These include aberrations at the chromosomal as well as the nucleotide level. The cytogenetic aberrations in male and female germ cells in all the stages of spermatogenesis and oogenesis include numerical abnormalities such as aneuploidy and structural aberrations.

Spermatogenesis continues from youth to old age, while oogenesis occurs from youth to the menopause, hence there is more chance for accumulation of DNA damage before conception through mutagenic/carcinogenic exposures in male than in female gametes.[1] In the female, damage is detected predominantly during meiosis in the oocyte stage and in the male, the spermatocytes are also used to evaluate chromosomal damage, as reviewed previously.[2] The presence of chromosomal aberrations, however, indicates a potential to induce more subtle and possibly transmissible chromosomal damage which survives cell

Issues in Toxicology No. 10
Biomarkers and Human Biomonitoring, Volume 2: Selected Biomarkers of Current Interest
Edited by Lisbeth E. Knudsen and Domenico Franco Merlo

Published by the Royal Society of Chemistry, www.rsc.org

division, producing heritable cytogenetic changes. Cytogenetic damage is usually accompanied by other genotoxic damage such as gene mutation.

Structural and numerical chromosomal aberrations in somatic cells are involved in the aetiology of neoplasia;[3] while in germ cells they can lead to perinatal mortality, dominant lethality or congenital malformations in the offspring[4–6] and some congenital tumours.[6,7] In a large investigation in two laboratories, where dominant lethal animal data were combined, involving around 7000 animals, in 1981, Anderson *et al.*[4] showed that early deaths per total implants in CD1-mice did not increase in a linear fashion as was anticipated. Instead, it was found that for control animals values increased between one and four implants and then plateaued to 15 implants, after which there was a further increase. This pattern was also echoed by the treatment groups. This suggests that the female exerts some physiological protective effect on the developing fetus (Figure 12.1).

Conventional mating methods in reproductive toxicology demand large animal numbers and are very expensive, and there has been a move away from such studies to use alternative methods which need significantly fewer animals or *in vitro* studies. These include the use of various biomarkers, such as highly sensitive polymerase chain reaction (PCR) methods using only a few animals or the Comet assay *in vitro*.

12.2 Spermatogenesis and Oogenesis

Spermatogenesis refers to a highly complex system of differentiation regulated via hormones, as well as cell to cell communication between germ and Sertoli

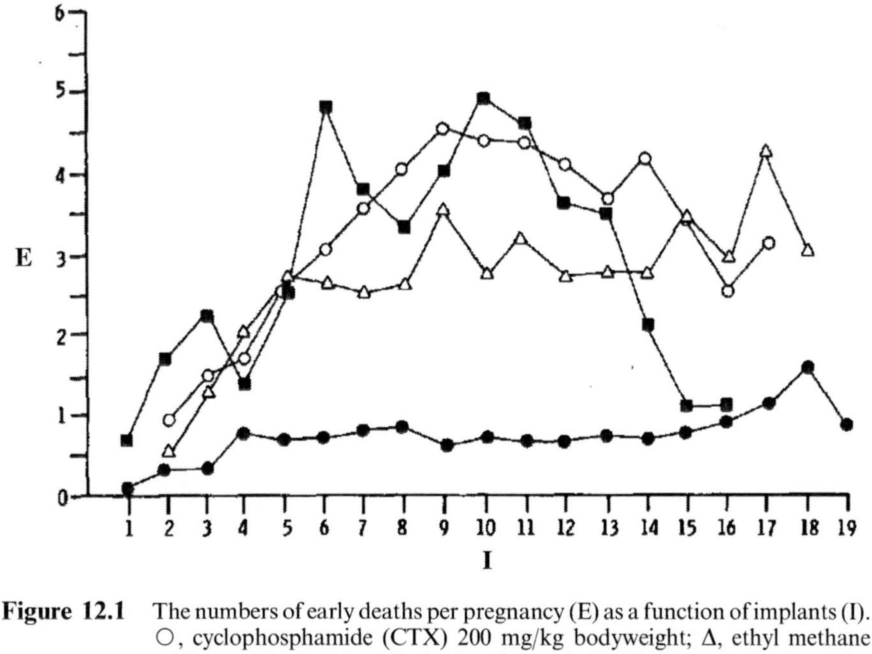

Figure 12.1 The numbers of early deaths per pregnancy (E) as a function of implants (I). ○, cyclophosphamide (CTX) 200 mg/kg bodyweight; Δ, ethyl methane sulfonate (EMS) 100 mg/kg bodyweight; ■, EMS 150 mg/kg bodyweight; ●, negative control.[4]

cells, to create mature sperm cells. Spermatogenesis is highly conserved in all mammalian species and occurs in three phases in the seminiferous tubules of the testis. During the proliferative phase (9 weeks in humans), primordial germ cells undergo first multiple mitotic divisions to form gonocytes, the stem cell for spermatogenesis. During human pre-puberty, these gonocytes differentiate to spermatogonia, which proliferate in many mitotic divisions. With the beginning of puberty, some spermatogonia differentiate to meiotic, primary spermatocytes (2n;2c). Note that the ploidy status is not doubled. These are 2n cells with doubled chromatids (2c). The number of centromeres defines the ploidy status and there are still 46 centromeres. Other cells continue to proliferate in mitotic divisions, representing a lifelong reservoir of stem cells for spermatogenesis. During the meiotic phase (*e.g.* 2 weeks in humans), the primary spermatocytes (2n;2c), which contain twice the DNA amount of a somatic cell, first divide during meiosis to form secondary spermatocytes (1n;2c) and then again during meiosis to form haploid spermatids (1n;1c). During spermiogenesis (*e.g.* 5 weeks in humans), the haploid spermatids undergo a fundamental biochemical and structural metamorphosis finally to form mature spermatozoa, which are then released from their connection to the Sertoli cells into the seminiferous tubules and are consequently pushed by hydrostatic pressure into the rete testis.[2]

The metamorphosis during the post-meiotic window involves the extrusion of a sperm tail, capture of the sperm head with an acrosome, creation of an energy transducing neck consisting of highly ordered mitochondria and finally the repackaging of the chromatin, to form a head-like structure. This repackaging of spermatid chromatin into six-fold condensed mature sperm chromatin represents a highly orchestrated remodelling process, in which spontaneous or induced mistakes can easily occur that result in transcriptionally inactive and consequently disabled (for the transcription of repair proteins) DNA sequences.[8,9]

The differentiation of the ovum, oogenesis, is different from spermatogenesis in many ways. The male germ cell, which originates in spermatogenesis, is essentially a motile nucleus, whereas the female germ cell formed during oogenesis has all that is necessary to initiate development and maintain metabolism. Beside the formation of a haploid nucleus during oogenesis, cytoplasmic enzymes, RNA, cellular structures and metabolic substrates are built up, making a complex cytoplasm, which is not the case in spermatozoa. However, the transcriptionally and translationally silent male germ cells are differentiated for motility.[1,10,11]

In addition, the mechanisms of oogenesis vary significantly in different species, in contrast to spermatogenesis. This is because oogenesis in many species, such as urchins and frogs, produces between hundreds and thousands of eggs, while most mammals produce far fewer eggs and humans generally only one or two. Where many ova are produced the oogonia represent self-renewing stem cells. In species producing fewer eggs a limited number of egg-precursor cells are formed. As an example, in the human up to the seventh month of embryonic development many millions of germ cells are produced. After that time the number of germ cells falls dramatically and many oogonia die, while the remaining oogonia proceed to the first meiotic division, and are termed primary oocytes. They progress through the first meiotic prophase and remain in this state until the diplotene stage when the female reaches puberty. After adolescence some of the eggs may continue to divide meiotically. Menstruation usually occurs around 11 years later and oocytes can be maintained in meiotic prophase for up to 54 years, when menopause occurs.[12,13]

In addition, the position of the metaphase plate during oogenic meiosis differs significantly from meiosis during spermatogenesis. During the division of the primary oocyte (2n;2c), the germinal vesicle (oocyte nucleus) collapses, allowing the metaphase spindle to migrate to the nuclear periphery. This results at telophase in one cell which contains minimal cytoplasm (termed the first polar body), whereas the other daughter cell has almost the total volume of normal cellular components and is termed the secondary oocyte (1n;2c). After meiosis I the chromosome set is reduced to n (1n) but with duplicated chromosomes (2c). Thus, one 1n1c ootide and 3 polar bodies are formed after meiosis II. The number of centromeres defines the ploidy status. A similar cytokinesis occurs during the second meiotic division, when most of the cytoplasm remains within the mature ovum (1n;1c) while a second polar body contains only very little more than the haploid nucleus. Therefore, only one of the progeny retains the full oocyte cytoplasm, whereas the other three do not (Figure 12.2).[12]

12.3 Biomarkers for Spermatozoa and Spermatocytes

A biomarker is an attribute or criterion that is objectively measured and evaluated as an indicator of certain processes, such as DNA damage or the aneuploidy status of cells. Here we describe the most common methodologies used as biomarkers of exposure, susceptibility or effect.

12.3.1 Introduction

Over 80% of all structural aberrations in animal studies have been shown to be of paternal origin,[14,15] whereas numerical aberrations are predominantly maternal in origin. Assays for spermatozoa, such as the use of fluorescence *in-situ* hybridization (FISH) assays, commonly detect possible heritable genetic diseases in predisposed individuals. Changes in morphology as well as changes in DNA integrity, measured in the Comet assay, have been shown to be correlated with infertility in many clinical studies. Assays for spermatozoa were also used to show significant differences in smokers and non-smokers when detecting DNA integrity, mRNA levels and DNA adduct formation.[14,16–18] A diagrammatic representation of methodologies to detect biomarkers in spermatozoa is shown in Figure 12.3.

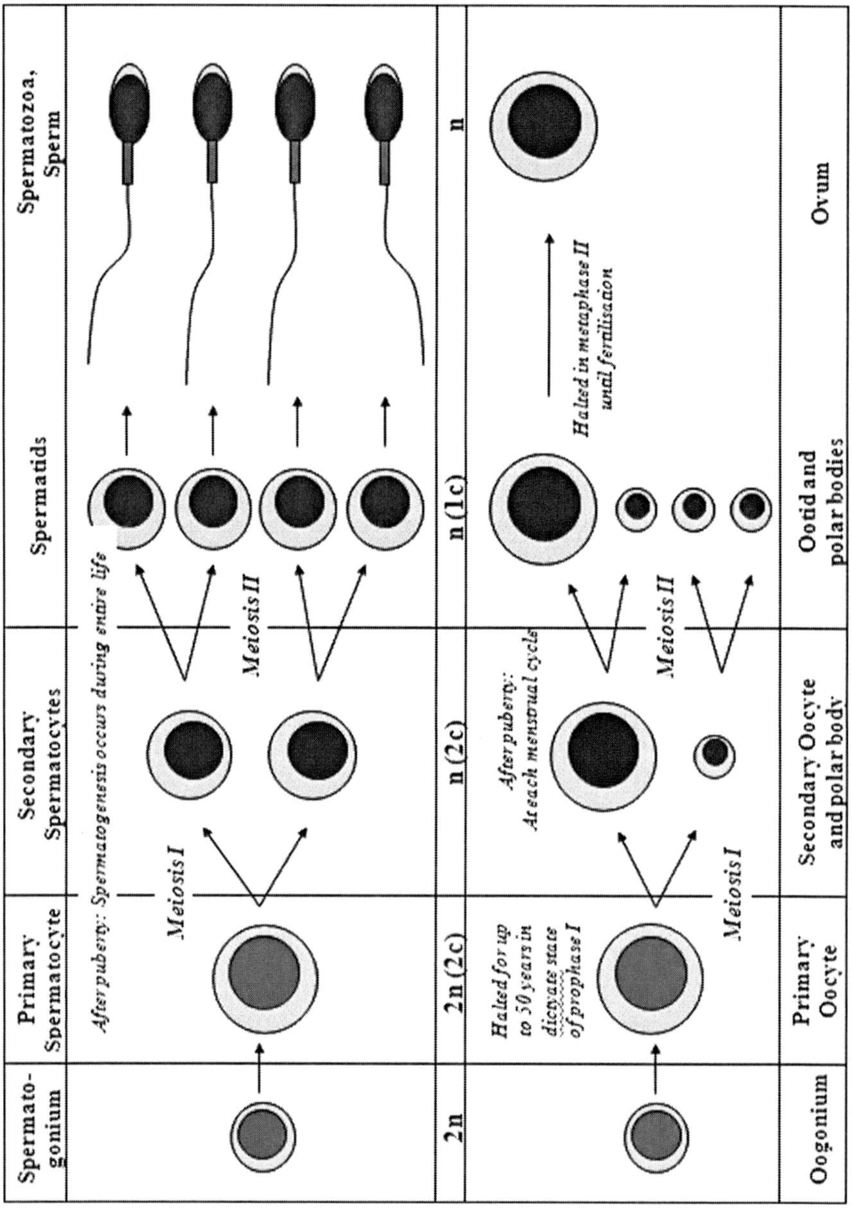

Figure 12.2 Comparison between spermatogenesis and oogenesis.

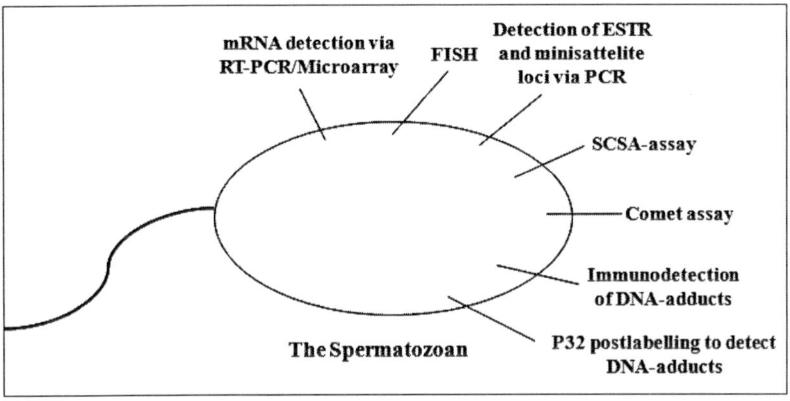

Figure 12.3 Common methodologies used as biomarkers in spermatozoa. Abbreviations: reverse transcriptase PCR (RT-PCR); fluorescence *in situ* hydridisation (FISH); sperm chromatin structure assay (SCSA); expanded single tandem repeat (ESTR).

12.3.2 Morphological Biomarkers for Spermatozoa

Animals can be treated *in vivo* with various chemicals to produce changes in sperm morphology; these changes without further treatment can be transmitted to the next generation.[19] In humans such changes have also been observed alongside chromosomal damage and DNA integrity changes following chemical exposure at the workplace or due to lifestyle habits.[20,21]

12.3.3 *In vivo* Comet Assay for Spermatozoa

The Comet assay is one of the most well-established biomarker systems for *in vivo* biomonotoring of occupational exposures and can be used in germ cells as well as somatic cells. Thus the assay will be considered in some detail. It is able rapidly and sensitively to test DNA-damaging genotoxins and confounding factors that influence responses.[22] The alkaline version of the Comet assay, with a pH ≥ 13, has become a reliable and an accepted assay for *in vivo* genotoxicological evaluations and has been approved by the UK Committee on Mutagenicity of Chemicals in Food, Consumer Products and Environment and the US Food and Drug Administration,[23] owing to the development of standardized protocols.[24–27] However, these guidelines are not entirely applicable when investigating reproductive cells such as spermatozoa in the Comet assay, unless several adjustments are made, particularly to relax the highly compacted sperm chromatin structure. In the regulatory setting, the Comet assay is primarily employed as a very useful follow-up or supplementary *in vivo* test for mutagenic compounds, which have been shown to produce a positive response in mutagenicity tests *in vitro* and a negative response in the bone marrow assay, because it demonstrates several advantages over other *in vivo* indicator tests that are generally accepted. The Comet assay can be applied to virtually any organ and differentiated cell type (local genotoxicity), provided an acceptable and suitable cell preparation method exists, and it covers a broader spectrum of primary DNA lesions by evaluating single cells.[28] It has been recommended[25] that 100 to 150 cells per individual have to be evaluated in the *in vivo* Comet assay applications, depending on the number of individuals per group. At least two dose levels are required to be tested: a high dose, which produces signs of toxicity, and a low dose (25–50% of the high dose). Owing to undergoing rapid DNA repair in cells other than mature sperm, *e.g.* spermatocytes, single strand breaks of primary lesions may only be short-lived. Therefore this kind of DNA damage could be missed by inadequate sampling times.[28]

Various studies facilitating the *in vivo* Comet assay on sperm or testicular cells have been done to toxicologically evaluate reprotoxins and genotoxins [see Table 12.1 in Baumgartner *et al.* for a complete overview].[29] A variety of toxicants has been investigated *in vivo* in mice including vanadium,[30,31] herbicides such as bentazon,[32] and X-rays.[33–36] In rats, chemotherapeutic drugs such as cyclophosphamide[37,38] and bleomycin either on its own[37] or in combination with etoposide and cis-platin[39] have been tested with the Comet assay on testicular cells and sperm. Also, chemicals such as ethyl methanesulfonate and the testicular toxin ethylene glycol monomethyl ether[37] have been examined. The *in vivo* Comet assay has also been used with human sperm for evaluating DNA sperm damage in occupational exposure of workers to toxicants such as acrylonitrile,[40] phthalates[41] and pesticides such as fenvalerate,[42] and in monitoring populations for environmental exposure to carbaryl and chlorpyrifos, both pesticides, which appear to be associated with increased DNA damage in human sperm.[43] When monitoring populations it became evident that a positive correlation between age, as well as caffeine intake, and DNA damage could be observed in sperm.[44] Regression analysis showed that DNA damage was positively associated with age (29–44 years), abnormal sperm and motility, and negatively associated with sperm concentration.[45]

It has been long known that the baseline DNA damage in human and mouse sperm in the Comet assay is high when compared to somatic cells, owing to the presence of alkali-labile sites.[46] Also, ejaculated sperm DNA is significantly more damaged than testicular sperm DNA.[47] Studies comparing baseline DNA damage in sperm from normozoospermic fertile, normozoospermic infertile and asthenozoospermic infertile groups did not show significant differences among the three groups. However, after challenge with X-rays and hydrogen peroxide it was concluded that the asthenozoospermic infertile group was more

Table 12.1 The cumulus–oocyte complex grading system, adapted from Wolf *et al.*[100]

Grade	*Morphological description*
Grade 1 (A)	Absent to sparse cumulus cells and 1–3 layers of corona cells
Grade 2 (B)	Dense cumulus cells and tightly packed corona cells
Grade 3 (C)	Expanded, fluffy cumulus cells and expanded corona cells
Grade 4 (D)	Expanded, scanty cumulus cells and expanded, often partially lost corona cells

susceptible to damage than the normozoospermic infertile group, which in turn was more susceptible than the fertile group. The fertile group contained a resistant subpopulation of spermatozoa with relatively intact DNA.[48,49] Irvine *et al.*[50] stated that a significant proportion of infertile men have elevated levels of DNA damage in their ejaculated spermatozoa. Highly significant negative correlations were observed between DNA fragmentation and semen quality, particularly sperm concentration. In addition, multiple regression analysis indicated that other attributes of semen quality, such as sperm movement and ROS generation, were also related to DNA damage.[50] Verit *et al.* did not find any relationship between sperm DNA damage and oxidative stress in normozoospermic infertile men and considered that the pathophysiology of idiopathic infertility cannot be explained by sperm DNA damage or seminal oxidative stress.[51] Trisini *et al.*[52] attempted to find associations between semen parameters and sperm DNA damage with the neutral Comet assay (Comet extent and tail moment). Although there were associations between semen and Comet assay parameters, their magnitudes were weak, suggesting that the Comet assay provides additional independent information on sperm function.[52]

12.3.4 *In vitro* Comet Assay for Spermatozoa

Studies with the *in vitro* Comet assay on sperm (see Table 12.2 below for a complete overview) have mainly been focused on the investigation of the potential genotoxic damage caused by compounds such as flavonoids (silymarin, myricetin, quercetin, kaempferol, rutin and kaempferol-3-rutinoside) and food mutagens [3-amino-1-methyl-5H-pyrido (4,3-b)indole (Trp) and 2-amino-3-methylimidazo-(4,5-f)quinoline (IQ)] either on their own or in

Table 12.2 The oocyte maturity grading system, adapted from Lin *et al.*[99]

Grade	*Morphological description*
Grade 1 (mature or pre-ovulatory)	Expanded cumulus, very radiant corona, distinct zona pellucida, clear ooplasm, expanded well aggregated membrana granulosa cells
Grade 2 (approximately mature)	Expanded cumulus mass, slightly compact corona radiata, expanded well aggregated membrane granulosa cells
Grade 3 (immature)	Dense compact cumulus if present, very adherent compact layer of corona cells, ooplasm if visible with the presence of the germinal vesicle, compact and non-aggregated membrana granulosa cells
Grade 4 (post-mature)	Much expanded cumulus with clumps, radiant corona radiata yet often clumped, irregular or incomplete, very visible zona, slightly granular or dark ooplasm, small and relatively non-aggregated membrana granulosa cells
Grade 5 (atretic)	Rarely with associated cumulus mass, clumped and very irregular corona radiata if present, very visible zona, dark and frequently misshapen ooplasm, membrana granulosa cells with very small clumps of cells

combination.[53] Further research has been carried out on oestrogens (diethylstilboestrol, beta-oestradiol, daidzein, genestein, equol and nonylphenyl) either on their own,[54] combined with antioxidants [catalase, vitamin C, superoxide dismutase (SOD)][55] or combined with flavonoids (quercetin, kaempferol).[56] Other toxicants investigated included X-rays,[57] gamma-radiation,[58] doxorubicin,[59] lead sulfate, nitrate and acetate, dibromochloropropane, ethylene glycol monoethyl ether, 1,2-epoxybutene, and 1,2,3,4-diepoxybutane.[54] All compounds produced positive responses, but ethylene glycol monoethyl ether only produced positive responses in sperm and not in peripheral lymphocytes, and similarly the phytoestrogens, genistein and daidzein, were less responsive in the peripheral lymphocytes in the male than in the sperm. This may be due to greater sensitivity of mature spermatozoa because of their lack of repair mechanisms.[54]

However, given that damage was generally seen over a similar dose range, a one-to-one or a one-to-two ratio of somatic and germ cell damage was observed and this has implications in humans for risk assessment purposes.[53,54] It was later concluded that human testicular cells have limited capacity to repair important oxidative DNA lesions, which could lead to impaired reproduction and *de novo* mutations.[60] By contrast, the usefulness of *in vitro* cultures of rat spermatocytes and Sertoli cells in conjunction with the Comet assay has been reported. This revealed the presence of DNA strand-breaks in non-treated cells, whose numbers decreased with the duration of the culture, suggesting the involvement of DNA repair mechanisms related to meiotic recombination.[61] Besides repair capacity, it should also be taken into account that when using cells from testes for *in vitro* studies, various testicular cell types show differences in metabolic activation of chemical compounds.[62] Anderson *et al.* believe that there are low levels of metabolic activity even in sperm because the heterocyclic amines that normally require metabolic activation have shown positive responses.[53]

12.3.5 Flow Cytometric Biomarkers for Spermatozoa

Flow cytometric analyses of specific and partly heterogeneous cell populations allow the quantitative evaluation of large numbers of cells in a very short time. Various cytometric assays that have been developed for somatic cells can also be employed in spermatoza to determine germ cell DNA damage accurately.

Flow cytometric analysis is based on the optical detection of single cells within a capillary passing though a focused laser beam. Parameters such as side-scattered and forward-scattered light describe the granularity and the size of each cell, respectively.[63] The sperm chromatin structure assay (SCSA) was the first cytometric assay using flow cytometry to evaluate the ratio of single- and double-strand breaks within individual spermatozoa.[64] Using the metachromatic dye acridine orange and the susceptibility of the sperm DNA to denaturation under acidic conditions, the SCSA assay has been shown to be highly dose-responsive in reproductive toxicology, while generating meaningful biological information on sperm DNA integrity.[65] Besides the SCSA assay, as one of the most frequently used flow-cytometric systems to assess sperm DNA fragmentation, the terminal deoxynucleotidyl transferase-mediated dUTP nick end-labelling (TUNEL) assay also has been widely employed.[66] As a biomarker for DNA integrity in somatic cells, the TUNEL assay allows, on the basis of the activation of endogenous nuclear endonucleases, the detection of the origin, the mechanism and the clinical meaning of sperm DNA fragmentation/damage in male germ cells.[67,68] By using fluorescein-labelled antibodies against the oxidative base adduct, 8-hydroxy-2′-deoxyguanosine (8OHdG) within sperm, De Iuliis and colleagues were able to correlate sperm DNA damage, evaluated by flow cytometry, with chromatin remodelling and the formation of 8OHdG, a biomarker of oxidative stress.[69]

Despite the analysis of various parameters on a single cell level and the useful possibility of cell sorting with subsequent assessment of different cell populations, flow cytometry cannot image a specific cell *in situ*. To overcome this problem, image cytometry such as laser-scanning cytometry (LSC) has been introduced.[70] This technique allows the rapid automated quantitative analysis of fluorescent signals of cells spread out on microscopic slides. As a biomarker of sperm DNA damage caused by exposure to an aneugenic toxicant, the induction of missegregated chromosomes can be reliably determined using LSC. After cytometric evaluation, image analysis allows the successful retrieval of abnormal cells, *e.g.* disomic sperm, directly on the slide, combining the virtues of fast image cytometry and microscopic FISH analysis.[71]

The plethora of possibilities is huge when increasing the number of lasers with different excitation wavelengths and, in succession, the number of fluorescent dyes. This allows for the fast simultaneous evaluation of more than one parameter, even further increasing the advantage of rapid delineation of various biomarkers of DNA damage in germ cells when compared to manual microscopic analysis.

12.3.6 Fluorescence *in situ* Hybridization (FISH) for Spermatozoa

Fluorescence *in situ* hybridization (FISH) was first developed for the visualization of nucleic acid sequences with radio-labelled probes in 1969,[72] and in 1980[73] it was replaced by the first *in situ* detection of sequences with fluorochrome-labelled probes, which is today's commonly used procedure for localization of specific nucleic acid sequences inside cells or tissues. In a typical FISH procedure, non-isotopically labelled DNA molecules ('probes') are incubated with cells or tissue sections which have been denatured to allow binding of probes to their cellular targets. Following overnight hybridization, excess probe molecules are removed by repeated wash steps so that only specifically bound probes can be seen in the fluorescence microscope. The efficiency of FISH, and thus the ability to detect a specific target inside a cell nucleus, depends on the penetration of probes and detection reagents as well as the accessibility of the hybridization target. FISH analysis of spermatozoa (termed 'sperm-FISH') can provide important information about structural or numerical chromosomal abnormalities of individuals suffering from infertility or possible severe genetic diseases.[74–77] In recent years, FISH on spermatozoa as well as on cells of pre-implantation embryos has been frequently used in countries with the appropriate legislation (such as the UK and USA) to determine possible genetic diseases in the offspring before fertilization/pre-implantation.[78–81]

12.3.7 Immunodetection or ^{32}P Labelling for DNA-adduct Detection as a Biomarker

Monoclonal antibodies raised against DNA adducts were used successfully to detect benzo(a)pyrene diol-epoxide DNA adducts and diol-epoxide DNA adducts of different polyaromatic hydrocarbons in a cross reaction in spermatozoa,[18,82] showing that these adducts are able to cross the blood–testis barrier. There are many other monoclonal antibodies against toxin induced DNA adducts which have been applied to somatic cells and might also work on spermatozoa, such as 2-amino-1-methyl-6-phenylimidazo[4,5–b]pyridine (PhIP) DNA adducts[83] or aflatoxin DNA adducts.[84] Similar results were obtained in spermatozoa using ^{32}P-postlabelling technology.

12.3.8 mRNA Profiles as Biomarkers for Spermatozoa

Because the testis can only be used in animal studies as a routine source for mRNA biomonitoring, human spermatozoa were shown to be a suitable alternative, which reflects a testicular response to gentotoxic exposures. This is because spermiogenesis and sperm maturation in the epididymis last for approximately 45–50 days in men. Linschooten *et al.* also showed, using reverse transcriptase-PCR (RT-PCR) and microarray gene expression, differences in the spermatozoa of smokers and non-smokers.[17]

12.3.9 Mutation Detection at ESTR and Minisatellite Loci for Spermatozoa

Detection of induced mutations at expanded tandem repeats (ESTR) or minisatellite loci via highly sensitive PCR methods showed significant increases in germ-line mutations of animals and humans exposed to radiation as well as various toxins [reviewed by Verhofstad *et al.* and Yauk].[85,86] These types of mutation are shown as the gain or loss of certain repetitive sequences detected by either amplified spermatozoa DNA or via screening of the offspring through pedigree analysis.[85,86] Dubrova *et al.* were able to show for the first time with this PCR methodology that the rate of germline mutations at repetitive sequences responded significantly to gamma-radiation exposure.[87]

12.3.10 Heritable Chromosome Assays for Spermatocytes

Damage may be analysed in the heritable translocation test, which involves the examination in male F_1 animals of diakinesis metaphase I spermatocytes for multivalent association.[88–90] Fertilized ova can also be analysed by cytogenetic analysis,[91] or early embryos can be examined by metaphase analysis.[92] Such techniques are technically demanding, and the heritable translocation assay requires large numbers of animals to attain appropriate sample sizes. Chromosome damage in fertilized ova or early embryos may not be compatible with survival after birth, so it is only the heritable translocation assay that provides absolute evidence of induced heritable effects. In fetal mice which have been exposed transplacentally after exposure of the mother, tissues can be assessed for micronuclei or metaphase analysis. Cole and colleagues scored micronuclei in polychromatic erythrocytes in the liver or peripheral blood of fetal mice.[93] Such methods have been reviewed and are used to investigate factors which might affect embryos.[94]

12.4 Biomarkers for Oocytes

12.4.1 Introduction

As opposed to spermatozoa, high numbers of oocytes (>4000) carry a numerical aberration as detected by conventional karyotyping techniques.[95] It was determined from these data that 13.5% were hypoploid and 8.6% hyperploid. Therefore, it was estimated that up to 17.2% of all oocytes carry a numerical aberration and it was suggested that these numbers come mostly from older women and account for the majority of *in vitro* fertilization (IVF) failures.[96] While some studies from assisted reproduction provided evidence for this maternal age effect, others did not support this finding, which contrasts with the findings from trisomy data.[97] The overall estimates from studies of oocytes in IVF may be rather high and may not reflect the overall rate of chromosomal aberrations in oocytes of healthy, fertile young women. This is because the number of oocytes investigated in conventional cytogenetic studies is very low and oocytes are extremely heterogeneous cells with respect to their origin, age of the female, clinical history and hormonal stimulation before oocyte retrieval.[95] There are many biomarkers used for studies on oocytes; the most important are summarized in the text and in Figure 12.4.

12.4.2 Morphological Biomarkers for Oocytes

Before intracytoplasmatic sperm injection the removal of cumulus cells is necessary, after which oocyte abnormality rates have been found to be between 60 and 70%.[98] These authors have also suggested that certain morphological features could provide useful indicators of oocyte qualities, based on the grading systems of Lin *et al.* and Wolf (see Tables 12.1 and 12.2).[99,100] Cytoplasmic defects involving density variations, texture and viscosity can affect the likelihood

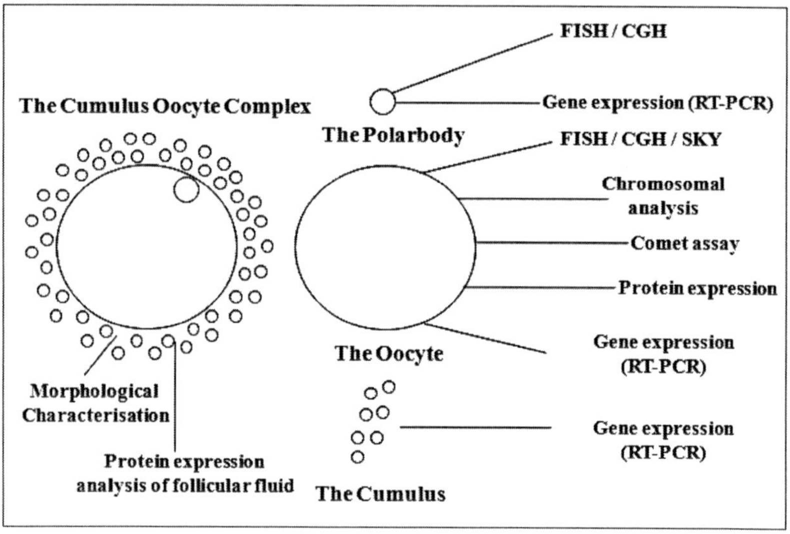

Figure 12.4 Common methodologies used as biomarkers in oocytes. Abbreviations: reverse transcriptase-PCR (RT-PCR); fluorescence *in situ* hybridisation (FISH); comparative genomic hybridisation (CGH); spectral karyotyping (SKY).

of a positive outcome following IVF treatment.[101] Vacuoles in the cytoplasm reduce fertility,[102] and blastocyst formation and pregnancy rates are affected by smooth endoplasmic reticulum aggregations.[103] The absence of metaphase 11 spindles is associated with reduced fertilization and blastocyst formation as well as a shift in the relative position of the spindle to the first polar body.[104]

12.4.3 Fluorescence *in situ* Hybridization (FISH) for Oocytes

For over a decade the identification of viable oocytes via screening for chromosomal anomalies with FISH assays has been recognized.[105] High rates of numerical aberrations are reported in the FISH assay using centromere and chromosomal-paint probes on human oocytes, including those from younger females,[106] in biopsies from first and second polar bodies[105,107–110] or in preimplantation embryos analysed for IVF.[111] For example FISH on oocyte biomarkers has detected in several studies a correlation between increases in aneuploidy and advanced maternal age.[109,112] Given that aneuploidy in metaphase II oocytes only reflects the extent of first meiotic errors, total aneuploidy levels appear to be very high in aged females. Hence, using conventional chromosomal analysis as well as FISH assays on preimplantation embryos it was shown that the aneuploidy levels in embryos of aged women reach up to 20% and that chromosomal malsegregation also increases significantly with maternal age.[113] It was also revealed that certain chromosomes can have differential aneuploidy levels, such as chromosome 16, which shows very high levels in aged women compared to other chromosomes.[114,115] It is well known that aneuploidy of chromosome 21 is involved in age-related non-disjunction of female meiosis, while X chromosomal errors, such as XO (Turner's syndrome) and XXY (Klinefelter's syndrome) are less frequent. It does appear that the high rates of numerical chromosomal aberrations in human oocytes, compared to the lower rates of aneuploidy in human spermatozoa, imply that maternal age and female meiosis contribute to failures in IVF schemes as well as to preimplantation losses, fertility reduction and aneuploidy in the human.[95]

12.4.4 Comparative Genomic Hybridization (CGH) for Oocytes

The use of CGH in order to quantify the relative copy number of chromosomes in polar bodies and oocytes has been only used in research studies, not in the clinic so far. It has the great advantage that all chromosomes are examined in one experiment, rather than a limited number of chromosomes as used with the FISH assay (conventionally up to three chromosomes).[116,117]

The obvious disadvantage of the CGH technology is its time-consuming analysis of the hybridization results, although recent advances in oocyte cryopreservation have accelerated the process. However, the introduction of microarray technologies, such as single nucleotide polymorphism (SNP) arrays or array CGH may overcome this limitation.[118,119]

12.4.5 Comet Assay for Oocytes

Given that oocytes do not have a protaminated chromatin structure, unlike spermatozoa, the conventional protocol for somatic cells as first developed by Ostling and Johanson can be used.[120] When they applied an electrophoretic field with pH below 10, 'tails' were observed where some DNA from the nucleus migrated faster than the rest of the nuclear DNA. Alkaline denaturation at a higher pH and DNA unwinding were incorporated later,[121] and seemed to be an important step in detecting DNA damage. This allowed, at a pH of ≥ 13, the detection of double-strand breaks (DSB), single-strand breaks (SSB) and alkali-labile sites (ALS). Under neutral conditions (pH 7–9) almost exclusively DSB can be detected by merely subjecting lysed cell nuclei to an electrophoretic field at neutral pH.[27,57,58] This highly reliable method was adopted by several groups to detect DNA integrity in oocytes, *e.g.* non-depellucidated and depellucidated examples among human, porcine, bovine or mouse oocytes.[122–125]

12.4.6 Gene Expression as a Biomarker for Oocytes

Using RT-PCR, various studies could show expression changes for specific genes related to quality of the oocyte or cumulus cells.[126–128] These findings suggested causative relationships between mRNA expression alterations, age of the female and aneuploidy,[129] as well as causative relationships between alterations in gene expression and abnormal preimplantation embryos.[128,130] However, this methodology is limited by the number of genes that can be assessed for each oocyte (<10 genes). Following RNA amplification high throughput technologies, such as microarrays, can overcome this limitation and many thousands of genes can be assessed in a single experiment using relatively less RNA by comparison with RT-PCR.[131]

Given that cumulus cells have the same follicular environment as the oocyte with which they are associated, analysis of gene expression in these cells may provide an indication of oocyte competence.[132,133]

12.4.7 Protein Expression as a Biomarker for Oocytes

There are only very few data addressing the protein expression patterns in the oocyte and so far there is no study that has found proteomic markers associated with viability, while being unrelated to morphology. However, analysis of the by-products of metabolism found in the oocyte media revealed data concerning oocyte viability and maturation; for example Roberts *et al.* (2004)[134] showed, *via* proteomic analysis, a dialogue between cumulus cells and the maturing oocyte that influences the follicle-stimulating hormone responsiveness and substrate metabolism of the whole cumulus–oocyte complex.[134] Also, there are few protein expression changes in the developing blastocysts by comparison with degenerating embryos.[135,136]

12.5 Conclusion

As has been reported above, there are many biomarkers currently available in male and female germ cells. Some are better established than others and some are newly developing. They show great promise for future research and have relevance for clinical use.

References

1. R. J. Aitken, P. Koopman and S. E. Lewis, Seeds of concern, *Nature*, 2004, **432**(7013), 48–52.
2. F. Marchetti and A. J. Wyrobek, Mechanisms and consequences of paternally-transmitted chromosomal abnormalities, *Birth Defects Res. C Embryo Today*, 2005, **75**(2), 112–129.
3. M. W. Beckmann, D. Niederacher, H. G. Schnurch, B. A. Gusterson and H. G. Bender, Multistep carcinogenesis of breast cancer and tumour heterogeneity, *J. Mol. Med.*, 1997, **75**(6), 429–439.
4. D. Anderson, D. B. McGregor and T. M. Weight, The relationship between early deaths and implants in control and mutagen-treated CD-1 mice in dominant lethal assays, *Mutat. Res.*, 1981, **81**(2), 187–196.
5. A. C. Chandley, The origin of chromosomal aberrations in man and their potential for survival and reproduction in the adult human population, *Ann. Genet.*, 1981, **24**(1), 5–11.

6. D. Anderson, P. C. Jenkinson, A. J. Edwards, J. A. Hughes and M. H. Brinkworth, Commentary: paternal legacies, *Teratog. Carcinog. Mutagen.*, 1999, **19**(2), 105–108.
7. T. Nomura, H. Nakajima, H. Ryo, L. Y. Li, Y. Fukudome, S. Adachi, H. Gotoh and H. Tanaka, Transgenerational transmission of radiation- and chemically induced tumors and congenital anomalies in mice: studies of their possible relationship to induced chromosomal and molecular changes, *Cytogenet. Genome Res.*, 2004, **104**(1–4), 252–260.
8. G. Boissonneault, Chromatin remodeling during spermiogenesis: a possible role for the transition proteins in DNA strand break repair, *FEBS Lett.*, 2002, **514**(2–3), 111–114.
9. A. L. Kierszenbaum and L. L. Tres, RNA transcription and chromatin structure during meiotic and postmeiotic stages of spermatogenesis, *Fed. Proc.*, 1978, **37**(11), 2512–2516.
10. S. Gilbert, in *Developmental Biology*, ed. S. Gilbert (Sinauer Associates, 2000), Vol. 6.
11. G. A. Sega, Unscheduled DNA synthesis (DNA repair) in the germ cells of male mice – its role in the study of mammalian mutagenesis, *Genetics*, 1979, **92**(1 Pt 1 Suppl), s49–58.
12. T. Baker, in *Germ cells and fertilization*, ed. A. Short (Camebridge University Press, 1982), pp. 17–45.
13. J. H. Pinkerton, K. D. Mc, E. C. Adams and A. T. Hertig, Development of the human ovary – a study using histochemical technics, *Obstet. Gynecol.*, 1961, **18**, 152–181.
14. A. C. Chandley, On the parental origin of de novo mutation in man, *J. Med. Genet.*, 1991, **28**(4), 217–223.
15. E. D. Sloter, X. Lowe, I. D. Moore, J. Nath and A. J. Wyrobek, Multicolor FISH analysis of chromosomal breaks, duplications, deletions, and numerical abnormalities in the sperm of healthy men, *Am. J. Hum. Genet.*, 2000, **67**(4), 862–872.
16. E. M. Berry, Sperm abnormalities and cigarette smoking, *Lancet*, 1981, **1**(8230), 1159.
17. J. O. Linschooten, F. J. Van Schooten, A. Baumgartner, E. Cemeli, J. Van Delft, D. Anderson and R. W. Godschalk, Use of spermatozoal mRNA profiles to study gene-environment interactions in human germ cells, *Mutat. Res.*, 2009, **667**(1–2), 70–76.
18. V. Sipinen, J. Laubenthal, A. Baumgartner, E. Cemeli, J. O. Linschooten, R. W. Godschalk, F. J. Van Schooten, D. Anderson and G. Brunborg, *In vitro* evaluation of baseline and induced DNA damage in human sperm exposed to benzo[a]pyrene or its metabolite benzo[a]pyrene-7,8-diol-9,10-epoxide, using the comet assay, *Mutagenesis*, 2010, **25**(4), 417–425.
19. S. A. Clode and D. Anderson, Germ and somatic cell abnormalities following *in vivo* administration of thymidine and adenine, *Mutat. Res.*, 1988, **200**(1–2), 249–254.
20. H. J. Evans, J. Fletcher, M. Torrance and T. B. Hargreave, Sperm abnormalities and cigarette smoking, *Lancet*, 1981, **1**(8221), 627–629.

21. A. J. Wyrobek, L. A. Gordon, J. G. Burkhart, M. W. Francis, R. W. Kapp, Jr., G. Letz, H. V. Malling, J. C. Topham and M. D. Whorton, An evaluation of the mouse sperm morphology test and other sperm tests in nonhuman mammals. A report of the U.S. Environmental Protection Agency Gene-Tox Program, *Mutat. Res.*, 1983, **115**(1), 1–72.
22. P. Moller, L. E. Knudsen, S. Loft and H. Wallin, The comet assay as a rapid test in biomonitoring occupational exposure to DNA-damaging agents and effect of confounding factors, *Cancer Epidemiol. Biomarkers Prev.*, 2000, **9**(10), 1005–1015.
23. B. Burlinson, R. R. Tice, G. Speit, E. Agurell, S. Y. Brendler-Schwaab, A. R. Collins, P. Escobar, M. Honma, T. S. Kumaravel, M. Nakajima, Y. F. Sasaki, V. Thybaud, Y. Uno, M. Vasquez and A. Hartmann, Fourth International Workgroup on Genotoxicity testing: results of the *in vivo* Comet assay workgroup, *Mutat. Res.*, 2007, **627**(1), 31–35.
24. R. R. Tice, E. Agurell, D. Anderson, B. Burlinson, A. Hartmann, H. Kobayashi, Y. Miyamae, E. Rojas, J. C. Ryu and Y. F. Sasaki, Single cell gel/comet assay: guidelines for *in vitro* and *in vivo* genetic toxicology testing, *Environ. Mol. Mutagen.*, 2000, **35**(3), 206–221.
25. A. Hartmann, E. Agurell, C. Beevers, S. Brendler-Schwaab, B. Burlinson, P. Clay, A. Collins, A. Smith, G. Speit, V. Thybaud and R. R. Tice, Recommendations for conducting the *in vivo* alkaline Comet assay. 4th International Comet Assay Workshop, *Mutagenesis*, 2003, **18**(1), 45–51.
26. V. J. McKelvey-Martin, M. H. Green, P. Schmezer, B. L. Pool-Zobel, M. P. De Meo and A. Collins, The single cell gel electrophoresis assay (comet assay): a European review, *Mutat. Res.*, 1993, **288**(1), 47–63.
27. A. R. Collins, The comet assay for DNA damage and repair: principles, applications, and limitations, *Mol. Biotechnol.*, 2004, **26**(3), 249–261.
28. S. Brendler-Schwaab, A. Hartmann, S. Pfuhler and G. Speit, The *in vivo* comet assay: use and status in genotoxicity testing, *Mutagenesis*, 2005, **20**(4), 245–254.
29. A. Baumgartner, E. Cemeli and D. Anderson, The comet assay in male reproductive toxicology, *Cell Biol. Toxicol.*, 2009, **25**(1), 81–98.
30. M. Altamirano-Lozano, L. Alvarez-Barrera, F. Basurto-Alcantara, M. Valverde and E. Rojas, Reprotoxic and genotoxic studies of vanadium pentoxide in male mice, *Teratog. Carcinog. Mutagen.*, 1996, **16**(1), 7–17.
31. P. Leopardi, P. Villani, E. Cordelli, E. Siniscalchi, E. Veschetti and R. Crebelli, Assessment of the *in vivo* genotoxicity of vanadate: analysis of micronuclei and DNA damage induced in mice by oral exposure, *Toxicol. Lett.*, 2005, **158**(1), 39–49.
32. S. Garagna, C. Vasco, V. Merico, A. Esposito, M. Zuccotti and C. A. Redi, Effects of a low dose of bentazon on spermatogenesis of mice exposed during foetal, postnatal and adult life, *Toxicology*, 2005, **212**(2–3), 165–174.
33. E. Cordelli, A. M. Fresegna, G. Leter, P. Eleuteri, M. Spano and P. Villani, Evaluation of DNA damage in different stages of mouse spermatogenesis after testicular X irradiation, *Radiat. Res.*, 2003, **160**(4), 443–451.

34. M. M. Dobrzynska, The effects in mice of combined treatments to X-rays and antineoplastic drugs in the Comet assay, *Toxicology*, 2005, **207**(2), 331–338.
35. G. A. Haines, J. H. Hendry, C. P. Daniel and I. D. Morris, Increased levels of comet-detected spermatozoa DNA damage following *in vivo* isotopic- or X-irradiation of spermatogonia, *Mutat. Res.*, 2001, **495**(1–2), 21–32.
36. G. A. Haines, J. H. Hendry, C. P. Daniel and I. D. Morris, Germ cell and dose-dependent DNA damage measured by the comet assay in murine spermatozoaa after testicular X-irradiation, *Biol. Reprod.*, 2002, **67**(3), 854–861.
37. D. Anderson, A. Dhawan, T. W. Yu and M. J. Plewa, An investigation of bone marrow and testicular cells *in vivo* using the comet assay, *Mutat. Res.*, 1996, **370**(3–4), 159–174.
38. A. M. Codrington, B. F. Hales and B. Robaire, Spermiogenic germ cell phase-specific DNA damage following cyclophosphamide exposure, *J. Androl.*, 2004, **25**(3), 354–362.
39. G. Delbes, B. F. Hales and B. Robaire, Effects of the chemotherapy cocktail used to treat testicular cancer on sperm chromatin integrity, *J. Androl.*, 2007, **28**(2), 241–249; discussion 250–241.
40. D. X. Xu, Q. X. Zhu, L. K. Zheng, Q. N. Wang, H. M. Shen, L. X. Deng and C. N. Ong, Exposure to acrylonitrile induced DNA strand breakage and sex chromosome aneuploidy in human spermatozoa, *Mutat. Res.*, 2003, **537**(1), 93–100.
41. S. M. Duty, N. P. Singh, M. J. Silva, D. B. Barr, J. W. Brock, L. Ryan, R. F. Herrick, D. C. Christiani and R. Hauser, The relationship between environmental exposures to phthalates and DNA damage in human sperm using the neutral comet assay, *Environ. Health Perspect.*, 2003, **111**(9), 1164–1169.
42. Q. Bian, L. C. Xu, S. L. Wang, Y. K. Xia, L. F. Tan, J. F. Chen, L. Song, H. C. Chang and X. R. Wang, Study on the relation between occupational fenvalerate exposure and spermatozoa DNA damage of pesticide factory workers, *Occup. Environ. Med.*, 2004, **61**(12), 999–1005.
43. J. D. Meeker, N. P. Singh, L. Ryan, S. M. Duty, D. B. Barr, R. F. Herrick, D. H. Bennett and R. Hauser, Urinary levels of insecticide metabolites and DNA damage in human sperm, *Hum. Reprod.*, 2004, **19**(11), 2573–2580.
44. T. E. Schmid, B. Eskenazi, A. Baumgartner, F. Marchetti, S. Young, R. Weldon, D. Anderson and A. J. Wyrobek, The effects of male age on sperm DNA damage in healthy non-smokers, *Hum. Reprod.*, 2007, **22**(1), 180–187.
45. I. D. Morris, S. Ilott, L. Dixon and D. R. Brison, The spectrum of DNA damage in human sperm assessed by single cell gel electrophoresis (Comet assay) and its relationship to fertilization and embryo development, *Hum. Reprod.*, 2002, **17**(4), 990–998.
46. N. P. Singh, D. B. Danner, R. R. Tice, M. T. McCoy, G. D. Collins and E. L. Schneider, Abundant alkali-sensitive sites in DNA of human and mouse sperm, *Exp. Cell Res.*, 1989, **184**(2), 461–470.

47. E. K. Steele, N. McClure and S. E. Lewis, Comparison of the effects of two methods of cryopreservation on testicular sperm DNA, *Fertil. Steril.*, 2000, **74**(3), 450–453.
48. C. M. Hughes, S. E. Lewis, V. J. McKelvey-Martin and W. Thompson, A comparison of baseline and induced DNA damage in human spermatozoa from fertile and infertile men, using a modified comet assay, *Mol. Hum. Reprod.*, 1996, **2**(8), 613–619.
49. V. J. McKelvey-Martin, N. Melia, I. K. Walsh, S. R. Johnston, C. M. Hughes, S. E. Lewis and W. Thompson, Two potential clinical applications of the alkaline single-cell gel electrophoresis assay: (1). Human bladder washings and transitional cell carcinoma of the bladder; and (2). Human sperm and male infertility, *Mutat. Res.*, 1997, **375**(2), 93–104.
50. D. S. Irvine, J. P. Twigg, E. L. Gordon, N. Fulton, P. A. Milne and R. J. Aitken, DNA integrity in human spermatozoa: relationships with semen quality, *J. Androl.*, 2000, **21**(1), 33–44.
51. F. F. Verit, A. Verit, A. Kocyigit, H. Ciftci, H. Celik and M. Koksal, No increase in sperm DNA damage and seminal oxidative stress in patients with idiopathic infertility, *Arch. Gynecol. Obstet.*, 2006, **274**(6), 339–344.
52. A. T. Trisini, N. P. Singh, S. M. Duty and R. Hauser, Relationship between human semen parameters and deoxyribonucleic acid damage assessed by the neutral comet assay, *Fertil. Steril.*, 2004, **82**(6), 1623–1632.
53. D. Anderson, N. Basaran, M. M. Dobrzynska, A. A. Basaran and T. W. Yu, Modulating effects of flavonoids on food mutagens in human blood and sperm samples in the comet assay, *Teratog. Carcinog. Mutagen.*, 1997, **17**(2), 45–58.
54. D. Anderson, M. M. Dobrzynska and N. Basaran, Effect of various genotoxins and reproductive toxins in human lymphocytes and sperm in the Comet assay, *Teratog. Carcinog. Mutagen.*, 1997, **17**(1), 29–43.
55. D. Anderson, T. E. Schmid, A. Baumgartner, E. Cemeli-Carratala, M. H. Brinkworth and J. M. Wood, Oestrogenic compounds and oxidative stress (in human sperm and lymphocytes in the Comet assay), *Mutat. Res.*, 2003, **544**(2–3), 173–178.
56. E. Cemeli, T. E. Schmid and D. Anderson, Modulation by flavonoids of DNA damage induced by estrogen-like compounds, *Environ. Mol. Mutagen.*, 2004, **44**(5), 420–426.
57. N. P. Singh and R. E. Stephens, X-ray induced DNA double-strand breaks in human sperm, *Mutagenesis*, 1998, **13**(1), 75–79.
58. G. Haines, B. Marples, P. Daniel and I. Morris, DNA damage in human and mouse spermatozoa after *in vitro*-irradiation assessed by the comet assay, *Adv. Exp. Med. Biol.*, 1998, **444**, 79–91; discussion 92–73.
59. A. Baumgartner, T. E. Schmid, E. Cemeli and D. Anderson, Parallel evaluation of doxorubicin-induced genetic damage in human lymphocytes and sperm using the comet assay and spectral karyotyping, *Mutagenesis*, 2004, **19**(4), 313–318.
60. A. K. Olsen, N. Duale, M. Bjoras, C. T. Larsen, R. Wiger, J. A. Holme, E. C. Seeberg and G. Brunborg, Limited repair of 8-hydroxy-7,8-

dihydroguanine residues in human testicular cells, *Nucleic Acids Res.*, 2003, **31**(4), 1351–1363.
61. J. Perrin, D. Lussato, M. De Meo, P. Durand, J. M. Grillo, M. R. Guichaoua, A. Botta and J. L. Berge-Lefranc, Evolution of DNA strand-breaks in cultured spermatocytes: the Comet Assay reveals differences in normal and gamma-irradiated germ cells, *Toxicol. In Vitro*, 2007, **21**(1), 81–89.
62. C. Bjorge, R. Wiger, J. A. Holme, G. Brunborg, R. Andersen, E. Dybing and E. J. Soderlund, *In vitro* toxicity of 1,2-dibromo-3-chloropropane (DBCP) in different testicular cell types from rats, *Reprod. Toxicol.*, 1995, **9**(5), 461–473.
63. U. Sack, in *Cellular Diagnostics* (Karger, Freiburg, 2009).
64. D. P. Evenson, Z. Darzynkiewicz and M. R. Melamed, Relation of mammalian sperm chromatin heterogeneity to fertility, *Science*, 1980, **210**(4474), 1131–1133.
65. D. P. Evenson, K. L. Larson and L. K. Jost, Sperm chromatin structure assay: its clinical use for detecting sperm DNA fragmentation in male infertility and comparisons with other techniques, *J. Androl.*, 2002, **23**(1), 25–43.
66. R. Henkel, C. F. Hoogendijk, P. J. Bouic and T. F. Kruger, TUNEL assay and SCSA determine different aspects of sperm DNA damage, *Andrologia*, 2010, **42**(5), 305–313.
67. W. Gorczyca, F. Traganos, H. Jesionowska and Z. Darzynkiewicz, Presence of DNA strand breaks and increased sensitivity of DNA *in situ* to denaturation in abnormal human sperm cells: analogy to apoptosis of somatic cells, *Exp. Cell Res.*, 1993, **207**(1), 202–205.
68. M. Muratori, L. Tamburrino, S. Marchiani, C. Guido, G. Forti and E. Baldi, Critical aspects of detection of sperm DNA fragmentation by TUNEL/flow cytometry, *Syst. Biol. Reprod. Med.*, 2010, **56**(4), 277–285.
69. G. N. De Iuliis, L. K. Thomson, L. A. Mitchell, J. M. Finnie, A. J. Koppers, A. Hedges, B. Nixon and R. J. Aitken, DNA damage in human spermatozoa is highly correlated with the efficiency of chromatin remodeling and the formation of 8-hydroxy-2′-deoxyguanosine, a marker of oxidative stress, *Biol. Reprod.*, 2009, **81**(3), 517–524.
70. L. A. Kamentsky and L. D. Kamentsky, Microscope-based multiparameter laser scanning cytometer yielding data comparable to flow cytometry data, *Cytometry*, 1991, **12**(5), 381–387.
71. A. Baumgartner, T. E. Schmid, H. K. Maerz, I. D. Adler, A. Tarnok and M. Nuesse, Automated evaluation of frequencies of aneuploid sperm by laser-scanning cytometry (LSC), *Cytometry*, 2001, **44**(2), 156–160.
72. J. G. Gall and M. L. Pardue, Formation and detection of RNA-DNA hybrid molecules in cytological preparations, *Proc. Natl. Acad. Sci. U. S. A.*, 1969, **63**(2), 378–383.
73. J. G. Bauman, J. Wiegant, P. Borst and P. van Duijn, A new method for fluorescence microscopical localization of specific DNA sequences by *in situ* hybridization of fluorochromelabelled RNA, *Exp. Cell Res.*, 1980, **128**(2), 485–490.

74. H. U. Weier, M. Wang, J. C. Mullikin, Y. Zhu, J. F. Cheng, K. M. Greulich, A. Bensimon and J. W. Gray, Quantitative DNA fiber mapping, *Hum. Mol. Genet.*, 1995, **4**(10), 1903–1910.
75. J. Fung, W. Hyun, P. Dandekar, R. A. Pedersen and H. U. Weier, Spectral imaging in preconception/preimplantation genetic diagnosis of aneuploidy: multicolor, multichromosome screening of single cells, *J. Assist. Reprod. Genet.*, 1998, **15**(5), 323–330.
76. J. Fung, H. U. Weier and R. A. Pedersen, Detection of structural and numerical chromosome abnormalities in interphase cells using spectral imaging, *J. Histochem. Cytochem.*, 2001, **49**(6), 797–798.
77. R. R. Swiger and J. D. Tucker, Fluorescence *in situ* hybridization: a brief review, *Environ. Mol. Mutagen.*, 1996, **27**(4), 245–254.
78. S. Munne, J. Fung, M. J. Cassel, C. Marquez and H. U. Weier, Preimplantation genetic analysis of translocations: case-specific probes for interphase cell analysis, *Hum. Genet.*, 1998, **102**(6), 663–674.
79. S. Munne, J. Grifo, J. Cohen and H. U. Weier, Chromosome abnormalities in human arrested preimplantation embryos: a multiple-probe FISH study, *Am. J. Hum. Genet.*, 1994, **55**(1), 150–159.
80. A. M. Estop, K. M. Cieply, S. Munne and E. Feingold, Multicolor fluorescence *in situ* hybridization analysis of the spermatozoa of a male heterozygous for a reciprocal translocation t(11;22)(q23;q11), *Hum. Genet.*, 1999, **104**(5), 412–417.
81. T. Escudero, I. Abdelhadi, M. Sandalinas and S. Munne, Predictive value of sperm fluorescence *in situ* hybridization analysis on the outcome of preimplantation genetic diagnosis for translocations, *Fertil. Steril.*, 2003, **79**(Suppl 3), 1528–1534.
82. M. T. Zenzes, R. Bielecki and T. E. Reed, Detection of benzo(a)pyrene diol epoxide-DNA adducts in sperm of men exposed to cigarette smoke, *Fertil. Steril.*, 1999, **72**(2), 330–335.
83. X. Ma, J. R. Idle, M. A. Malfatti, K. W. Krausz, D. W. Nebert, C. S. Chen, J. S. Felton, D. J. Waxman and F. J. Gonzalez, Mouse lung CYP1A1 catalyzes the metabolic activation of 2-amino-1-methyl-6-phenylimidazo[4,5-b]pyridine (PhIP), *Carcinogenesis*, 2007, **28**(3), 732–737.
84. Y. J. Zhang, C. J. Chen, C. S. Lee, B. Haghighi, G. Y. Yang, L. W. Wang, M. Feitelson and R. Santella, Aflatoxin B1-DNA adducts and hepatitis B virus antigens in hepatocellular carcinoma and non-tumorous liver tissue, *Carcinogenesis*, 1991, **12**(12), 2247–2252.
85. N. Verhofstad, J. O. Linschooten, J. van Benthem, Y. E. Dubrova, H. van Steeg, F. J. van Schooten and R. W. Godschalk, New methods for assessing male germ line mutations in humans and genetic risks in their offspring, *Mutagenesis*, 2008, **23**(4), 241–247.
86. C. L. Yauk, Advances in the application of germline tandem repeat instability for in situ monitoring, *Mutat. Res.*, 2004, **566**(2), 169–182.
87. Y. E. Dubrova, A. J. Jeffreys and A. M. Malashenko, Mouse minisatellite mutations induced by ionizing radiation, *Nat. Genet.*, 1993, **5**(1), 92–94.

88. B. M. Cattanach, The heritable translocation test in mice. *In Cytogenetic Assays of Environmental Mutagens*, Ed. T. S. Hsu, (Allanheld, Osmun and Co., Totowa, New Jersey, 1982).
89. A. Leonard, Observations on meiotic chromosomes of the male mouse as a test of the potential mutagenicity of chemicals. *In Chemical Mutagens. Principles and Methods for Their Detection,* Ed. A. Hollender, (Plenum Press, New York, 1973).
90. A. Leonard, Tests for heritable translocations in male mammals, *Mutat. Res.*, 1975, **31**(5), 291–298.
91. R. Albanese, Mammalian male germ cell cytogenetics, *Mutagenesis*, 1987, **2**(2), 79–85.
92. I. Hansmann, Induced chromosomal aberrations in pronuclei, 2-cell stages and morulae of mice, *Mutat. Res.*, 1973, **20**(3), 353–367.
93. R. J. Cole, N. Taylor, J. Cole and C. F. Arlett, Short-term tests for transplacentally active carcinogens. I. Micronucleus formation in fetal and maternal mouse erythroblasts, *Mutat. Res.*, 1981, **80**(1), 141–157.
94. L. Henderson, Transplacental genotoxic agents: cytogenetic methods for their detection. *In Chemical Mutagens. Principles and Methods for Their Detection*, Ed. F. J. de Serres, (Plenum Press, New York, 1986).
95. U. Eichenlaub-Ritter, Genetics of oocyte ageing, *Maturitas*, 1998, **30**(2), 143–169.
96. T. Hassold and D. Chiu, Maternal age-specific rates of numerical chromosome abnormalities with special reference to trisomy, *Hum. Genet.*, 1985, **70**(1), 11–17.
97. K. Imai, N. Inukai, T. Ishida, T. Ozasa and S. Kawahara, [Studies on the syntheses of sulfur containing pyrimido [5,4-d]pyrimidine derivatives. III. A new synthesis of dipyridamole], *Yakugaku Zasshi*, 1976, **96**(5), 593–599.
98. B. Balaban and B. Urman, Effect of oocyte morphology on embryo development and implantation, *Reprod. Biomed. Online*, 2006, **12**(5), 608–615.
99. Y. C. Lin, S. Y. Chang, K. C. Lan, H. W. Huang, C. Y. Chang, M. Y. Tsai, F. T. Kung and F. J. Huang, Human oocyte maturity *in vivo* determines the outcome of blastocyst development *in vitro*, *J. Assist. Reprod. Genet.*, 2003, **20**(12), 506–512.
100. D. Wolf, *Oocyte quality and fertilization* (Plenum Press, 1988).
101. T. Ebner, M. Moser, M. Sommergruber, M. Puchner, R. Wiesinger and G. Tews, Developmental competence of oocytes showing increased cytoplasmic viscosity, *Hum. Reprod.*, 2003, **18**(6), 1294–1298.
102. T. Ebner, M. Moser, M. Sommergruber, U. Gaiswinkler, O. Shebl, K. Jesacher and G. Tews, Occurrence and developmental consequences of vacuoles throughout preimplantation development, *Fertil. Steril.*, 2005, **83**(6), 1635–1640.
103. J. Otsuki, A. Okada, K. Morimoto, Y. Nagai and H. Kubo, The relationship between pregnancy outcome and smooth endoplasmic reticulum clusters in MII human oocytes, *Hum. Reprod.*, 2004, **19**(7), 1591–1597.
104. L. Rienzi, F. Ubaldi, F. Martinez, M. Iacobelli, M. G. Minasi, S. Ferrero, J. Tesarik and E. Greco, Relationship between meiotic spindle location

with regard to the polar body position and oocyte developmental potential after ICSI, *Hum. Reprod.*, 2003, **18**(6), 1289–1293.
105. Y. Verlinsky, J. Cieslak, V. Ivakhnenko, S. Evsikov, G. Wolf, M. White, A. Lifchez, B. Kaplan, J. Moise, J. Valle, N. Ginsberg, C. Strom and A. Kuliev, Preimplantation diagnosis of common aneuploidies by the first- and second-polar body FISH analysis, *J. Assist. Reprod. Genet.*, 1998, **15**(5), 285–289.
106. T. Dailey, B. Dale, J. Cohen and S. Munne, Association between non-disjunction and maternal age in meiosis-II human oocytes, *Am. J. Hum. Genet.*, 1996, **59**(1), 176–184.
107. M. Grossmann, J. M. Calafell, V. Moreno, J. Balasch, J. A. Vanrell, J. Egozcue, J. Santalo and F. Vidal, Recurrent *in vitro* fertilization failure evaluated by fluorescence in situ hybridization: a case report, *Fertil. Steril.*, 1998, **69**(3), 558–560.
108. E. Martini, S. P. Flaherty, N. J. Swann, D. Payne and C. D. Matthews, Analysis of unfertilized oocytes subjected to intracytoplasmic sperm injection using two rounds of fluorescence in-situ hybridization and probes to five chromosomes, *Hum. Reprod.*, 1997, **12**(9), 2011–2018.
109. S. Munne, T. Dailey, K. M. Sultan, J. Grifo and J. Cohen, The use of first polar bodies for preimplantation diagnosis of aneuploidy, *Hum. Reprod.*, 1995, **10**(4), 1014–1020.
110. Y. Verlinsky, J. Cieslak, V. Ivakhnenko, S. Evsikov, G. Wolf, M. White, A. Lifchez, B. Kaplan, J. Moise, J. Valle, N. Ginsberg, C. Strom and A. Kuliev, Chromosomal abnormalities in the first and second polar body, *Mol. Cell Endocrinol.*, 2001, **183**(Suppl 1), S47–49.
111. Y. Verlinsky, J. Cieslak and A. Kuliev, Preimplantation FISH diagnosis of aneuploidies, *Methods Mol. Biol.*, 2002, **204**, 259–273.
112. J. D. Delhanty, D. K. Griffin, A. H. Handyside, J. Harper, G. H. Atkinson, M. H. Pieters and R. M. Winston, Detection of aneuploidy and chromosomal mosaicism in human embryos during preimplantation sex determination by fluorescent in situ hybridisation, (FISH), *Hum. Mol. Genet.*, 1993, **2**(8), 1183–1185.
113. M. E. Jamieson, J. R. Coutts and J. M. Connor, The chromosome constitution of human preimplantation embryos fertilized *in vitro*, *Hum. Reprod.*, 1994, **9**(4), 709–715.
114. S. Munne, A. Lee, Z. Rosenwaks, J. Grifo and J. Cohen, Diagnosis of major chromosome aneuploidies in human preimplantation embryos, *Hum. Reprod.*, 1993, **8**(12), 2185–2191.
115. C. A. Benadiva, I. Kligman and S. Munne, Aneuploidy 16 in human embryos increases significantly with maternal age, *Fertil. Steril.*, 1996, **66**(2), 248–255.
116. E. Fragouli, D. Wells, A. Thornhill, P. Serhal, M. J. Faed, J. C. Harper and J. D. Delhanty, Comparative genomic hybridization analysis of human oocytes and polar bodies, *Hum. Reprod.*, 2006, **21**(9), 2319–2328.
117. D. Wells, T. Escudero, B. Levy, K. Hirschhorn, J. D. Delhanty and S. Munne, First clinical application of comparative genomic hybridization

and polar body testing for preimplantation genetic diagnosis of aneuploidy, *Fertil. Steril.*, 2002, **78**(3), 543–549.
118. V. Bianchi, G. Coticchio, V. Distratis, N. Di Giusto, C. Flamigni and A. Borini, Differential sucrose concentration during dehydration (0.2 mol/l) and rehydration (0.3 mol/l) increases the implantation rate of frozen human oocytes, *Reprod. Biomed. Online*, 2007, **14**(1), 64–71.
119. A. Borini, C. Lagalla, M. A. Bonu, V. Bianchi, C. Flamigni and G. Coticchio, Cumulative pregnancy rates resulting from the use of fresh and frozen oocytes: 7 years' experience, *Reprod. Biomed. Online*, 2006, **12**(4), 481–486.
120. O. Ostling and K. J. Johanson, Microelectrophoretic study of radiation-induced DNA damages in individual mammalian cells, *Biochem. Biophys. Res. Commun.*, 1984, **123**(1), 291–298.
121. N. P. Singh, M. T. McCoy, R. R. Tice and E. L. Schneider, A simple technique for quantitation of low levels of DNA damage in individual cells, *Exp. Cell Res.*, 1988, **175**(1), 184–191.
122. A. Berthelot-Ricou, J. Perrin, C. Di Giorgio, M. De Meo, A. Botta and B. Courbiere, Comet assay on mouse oocytes: an improved technique to evaluate genotoxic risk on female germ cells, *Fertil. Steril.*, 2010, in press.
123. B. Jebelli, P. J. Chan, J. Corselli, W. C. Patton and A. King, Oocyte comet assay of luteal phase sera from nonpregnant patients after assisted reproductive procedures, *J. Assist. Reprod. Genet.*, 2001, **18**(8), 421–425.
124. H. Men, R. L. Monson, J. J. Parrish and J. J. Rutledge, Detection of DNA damage in bovine metaphase II oocytes resulting from cryopreservation, *Mol. Reprod. Dev.*, 2003, **64**(2), 245–250.
125. B. Novotna, J. Petr, M. Sedmikova, J. Kratochvilova and F. Jilek, Effect of different activation modes on DNA integrity of porcine M II oocytes matured *in vitro*, *Zygote*, 2010, **18**(1), 81–87.
126. M. A. Dode, I. Dufort, L. Massicotte and M. A. Sirard, Quantitative expression of candidate genes for developmental competence in bovine two-cell embryos, *Mol. Reprod. Dev.*, 2006, **73**(3), 288–297.
127. D. F. Russell, S. Baqir, J. Bordignon and D. H. Betts, The impact of oocyte maturation media on early bovine embryonic development, *Mol. Reprod. Dev.*, 2006, **73**(10), 1255–1270.
128. D. Wells, M. G. Bermudez, N. Steuerwald, A. R. Thornhill, D. L. Walker, H. Malter, J. D. Delhanty and J. Cohen, Expression of genes regulating chromosome segregation, the cell cycle and apoptosis during human preimplantation development, *Hum. Reprod.*, 2005, **20**(5), 1339–1348.
129. N. Steuerwald, J. Cohen, R. J. Herrera, M. Sandalinas and C. A. Brenner, Association between spindle assembly checkpoint expression and maternal age in human oocytes, *Mol. Hum. Reprod.*, 2001, **7**(1), 49–55.
130. D. Wells, M. G. Bermudez, N. Steuerwald, H. E. Malter, A. R. Thornhill and J. Cohen, Association of abnormal morphology and altered gene expression in human preimplantation embryos, *Fertil. Steril.*, 2005, **84**(2), 343–355.
131. P. Patrizio, E. Fragouli, V. Bianchi, A. Borini and D. Wells, Molecular methods for selection of the ideal oocyte, *Reprod. Biomed. Online*, 2007, **15**(3), 346–353.

132. I. Hernandez-Gonzalez, I. Gonzalez-Robayna, M. Shimada, C. M. Wayne, S. A. Ochsner, L. White and J. S. Richards, Gene expression profiles of cumulus cell oocyte complexes during ovulation reveal cumulus cells express neuronal and immune-related genes: does this expand their role in the ovulation process?, *Mol. Endocrinol.*, 2006, **20**(6), 1300–1321.
133. M. Shimada, I. Hernandez-Gonzalez, I. Gonzalez-Robayna and J. S. Richards, Paracrine and autocrine regulation of epidermal growth factor-like factors in cumulus oocyte complexes and granulosa cells: key roles for prostaglandin synthase 2 and progesterone receptor, *Mol. Endocrinol.*, 2006, **20**(6), 1352–1365.
134. R. Roberts, J. Stark, A. Iatropoulou, D. L. Becker, S. Franks and K. Hardy, Energy substrate metabolism of mouse cumulus-oocyte complexes: response to follicle-stimulating hormone is mediated by the phosphatidylinositol 3-kinase pathway and is associated with oocyte maturation, *Biol. Reprod.*, 2004, **71**(1), 199–209.
135. M. G. Katz-Jaffe, D. K. Gardner and W. B. Schoolcraft, Proteomic analysis of individual human embryos to identify novel biomarkers of development and viability, *Fertil. Steril.*, 2006, **85**(1), 101–107.
136. M. G. Katz-Jaffe, W. B. Schoolcraft and D. K. Gardner, Analysis of protein expression (secretome) by human and mouse preimplantation embryos, *Fertil. Steril.*, 2006, **86**(3), 678–685.

CHAPTER 13

Biomarkers of Exposure and Effect: Ionizing Radiation

A. FUCIC*[,a] AND A. SOROKIN[b]

[a] Institute for Medical Research and Occupational Health, Zagreb, Croatia; [b] School of Informatics, University of Edinburgh, United Kingdom; Institute of Cell Biophysics RAS, Pushchino, Moscow region, 142290, Russia

13.1 Introduction

There are two basic types of ionizing radiation: fast moving particles originating from radionuclides, and high energy photons. In a natural environment they usually occur together. The distribution of ionizing and excitation events along the particle track is called linear energy transfer (LET), which is defined as the energy in electron volts delivered per micron of particle track in the irradiated molecular system. The demarcation value between low and high LET is at 10 keV/μm. Types and LET of ionizing radiation are presented in Table 13.1.

In addition to the particular type of ionizing radiation, the biological effect depends on the quantity of transferred energy, *i.e.* the LET. The effect of ionizing radiation on cells differs for alpha, beta, gamma, and X-rays and depends on the route of exposure (external and/or internal). Some elements such as uranium produce a dual biological effect, as a source of radiation and xenoestrogen. The effects of radiation on living systems are very complex. Accidental irradiation with high doses causes acute radiation syndrome and is difficult to treat. Overexposure to low doses and low dose rates may cause tumors and cardiovascular diseases. In human occupational and living environments, there exists a broad range of ionizing radiation sources such as clinical diagnostic methods, tobacco (polonium), nuclear plants, nuclear weapons, surface mines (uranium), the nuclear weapons industry, terrorist

Issues in Toxicology No. 10
Biomarkers and Human Biomonitoring, Volume 2: Selected Biomarkers of Current Interest
Edited by Lisbeth E. Knudsen and Domenico Franco Merlo

Published by the Royal Society of Chemistry, www.rsc.org

Table 13.1 Types of ionizing radiation according to linear energy transfer (LET)

	Photons	
low LET		*(keV/μm)*
X-rays	250 kV	2
gamma ^{60}Co	1.2 MeV	0.3
	Particles	
low LET		*(keV/μm)*
electrons	1 MeV	0.25
high LET		*(keV/μm)*
Neutron	14 MeV	12
Alpha	5 MeV	166
Fe ions	2 GeV	1000

attacks using dirty bombs, *etc.* These exposures are usually accompanied by other chemical xenobiotics, possibly acting either independently, synergistically or antagonistically with respect to the ionizing radiation.

The mechanisms involved in the interaction between ionizing radiation and a living organism include direct DNA molecule breakage, lipid peroxidation disturbances, production of free radical species, methylation disturbances, RNA signal disturbances, genome instability, and bystander effects. The response of a living organism to ionizing radiation varies significantly and depends on age and sex, although such data are seldom incorporated in biodosometry calculations.

Exposure to ionizing radiation can be monitored successfully by personal dosimeters. An occupationally exposed population is continuously monitored with adherence to strict limits of annual dose. However, this is not the case with the general population. A received dose resulting from overexposure in nuclear accidents is usually estimated by the use of biomarkers. The advantage of biomarkers is that they provide information on both the dose received and the magnitude of the biological effect. Nonetheless, information on long-term effects, the bystander effect or genome instability is rarely available for the population in general.

Most biomarkers of ionizing radiation are used to assess external exposure. On the contrary, in tissues that accumulate radioisotopes such as the lung, thyroid, breast, or bones, the most frequently used cytogenetical biomarkers may produce false negative results while dose–response curves are particularly difficult to obtain.

Although it is possible to use certain hematological biomarkers, hair follicles or biochemical biomarkers to assess the effects of ionizing radiation on health, the most frequently used methods are cytogenetic assays. Cytogenetic assays provide information on dose using dose–response curves and serve to assess the health risk. However, these methods do not distinguish between exposure to ionizing radiation and that to chemicals. In addition, most cytogenetic biomarkers are short-lived and cannot be used to measure lifetime exposure.

New genome- and proteome-wide tools show cell changes within minutes to hours after exposure to ionizing radiation. Large-scale changes occur in genome profiles, involving a broad variety of cellular pathways following exposure to a wide range of low-dose (<10 cGy) and high-dose (>10 cGy) ionizing radiation.

Biodosimetry of the fetus and the child is the most demanding procedure and is usually based on mathematical models of phantoms. The major problem with the application of biomarkers in children is sampling. Methods providing reliable results and requiring samples as small as 100 μL are mostly suitable for newborns and children.

For decades the reference values used in biomonitoring did not distinguish between sexes and age groups, and instead were based on the typical Caucasian adult male. Today, efforts are being made to shift risk assessment from a group to an individual level.

13.2 Cytogenetic Biomarkers of Ionizing Radiation

Somatic mutations have been used over the last few decades as biomarkers of various early biological effects after exposure to ionizing radiation. They include glycophorin A (GPA) variants in erythrocytes and mutations in human leukocyte antigen (HLA) or hypoxanthine-guanine phosphoribosyl transferase (HPRT) loci in T lymphocytes. The GPA assay is a reliable biomarker of accidental overexposure and is stable over time, because it is based on bone marrow hematopoietic progenitor cells. The disadvantage of this method is that it is not reliable for low-dose exposures. The HPRT assay shows significant increases in mutant cell frequencies in a dose-dependent manner, but the protocol of the method, such as time of sampling, may affect the final results.

Ionizing radiation induces chromosomal aberrations such as mutations, breaks, deletions, and translocations. These cytogenetic biomarkers are used to predict an increased health risk, in particular cancer.[1] Genome damage caused by ionizing radiation is most often measured in lymphocytes, which are considered natural dosimeters, because they circulate through the body and are easily cultivated in the laboratory. The basic form of genome damage used in biodosimetry and dose–response curves is the dicentric chromosome (Figure 13.1).

The dicentric chromosome is formed when two chromosome segments from two different chromosomes (chromatids) join. The new aberrant chromosome has two centromeres with acentric fragments which are lost in the micronuclei after the first cell division. However, when the chromosome breaks, acentric fragments not attached to any chromosome can be detected without dicentric formation. The chromosomes may, in the case of two chromosome breaks

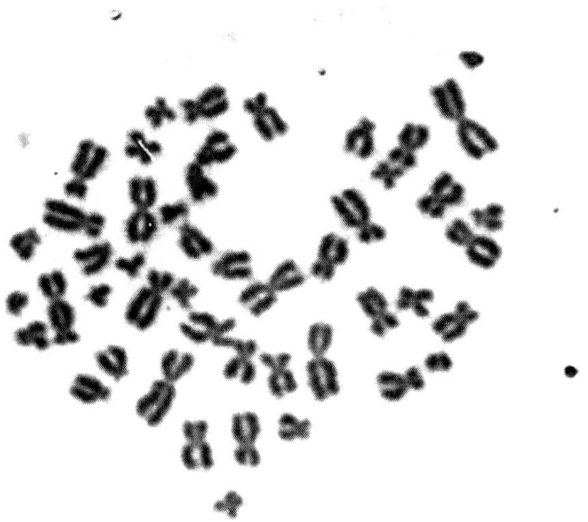

Figure 13.1 Chromosome aberration assay. Metaphase with dicentric chromosome.

within the same chromosome, join its own broken ends and form a ring chromosome. Exposure to high doses of ionizing radiation can produce a number of dicentric or even tricentric chromosomes in a cell, and the cell is therefore called multiaberrant. Multiaberrant cells are more often identified following internal exposure to radioisotopes. Furthermore, ionizing radiation can cause deletions, that is, the loss of part of the DNA in a chromosome, leading sometimes to the loss of tumor suppressor genes. Even though dicentrics, acentric fragments, chromosome and chromatid breaks are reliable predictors of an increased cancer risk, they can only detect recent exposure, say within a year of sampling, and cannot detect the cumulative effects of long-term exposure.

A crucial improvement in biodosimetry has come from fluorescent *in situ* hybridization (FISH), which possesses the ability to detect permanent chromosome rearrangements, translocations, and inversions. A translocation is formed if an acentric fragment joins a broken chromatid. If a broken chromatid reattaches to the same chromosome, it becomes inverted and may activate oncogenes and fuse genes. In most biodosimetry studies based on FISH, the genome equivalent[2] is calculated on a selected set of chromosomes. Most often, FISH is performed on chomosomes 1, 2 and 4 and genome damage is recalculated for the rest, assuming random genome damage on all chromosomes (Figure 13.2). A more accurate but less commonly available version is multicolor FISH, which enables analysis of all chromosomes.

FISH of metaphase chromosomes in suspension has resolved issues relating to scorer bias, slide quality and low sample cell number.[3,4] It allows rapid counting of painted human chromosomes and fast detection of chromosomal exchanges and rearrangements in hundreds of people a week.

Reliable dose–effect curves based on chromosome rearrangements have been developed and may be used in cases of mass radiation overexposure.[5] While a

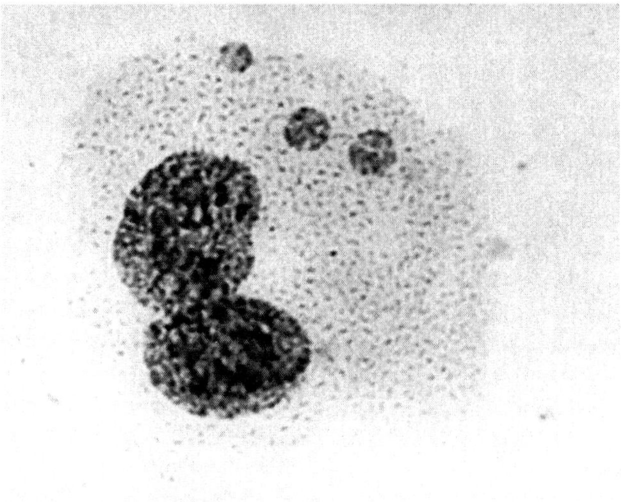

Figure 13.2 Binucleated lymphocyte with micronuclei.

dicentric chromosome is lethal for the cell, translocations can persist for decades. Cells with translocations are involved in carcinogenesis and genetic disorders. These methods can be used in the dose range between 0.2 and 4.0 Gy of low LET radiation. Future application of FISH will include prediction of the elimination of genome damage and estimating health risks on an individual basis.

Recent studies have shown that an increased risk of breast cancer in women occupationally exposed to ionizing radiation may be related to polymorphisms in chromosome 5p12 in the mitochondrial ribosomal protein S30.[6] It is interesting to note that in the Black Women's Health Study, the same chromosome 5p12 significantly increased breast cancer risk.[7] This example shows that chromosome rearrangements can be used in combination with proteomics and may contribute significantly to individual biomonitoring and risk assessment.

The second most common cytogenetic method is the micronuclues assay, which enables estimation of genome damage caused by both clastogenic and aneugenic mechanisms (Figure 13.3). The micronucleus assay may also be used in measuring proliferation and apoptosis and is a reliable predictor of increased cancer risk.[8] A micronucleus can contain a part of or a whole chromosome, which is excluded from the cell nucleus in the form of acentric fragment, dicentric chromosome, ring chromosome, or undamaged chromosome (as a result of a disturbance of the mitotic spindle). Although the aneugenic effect of ionizing radiation on the structure of the mitotic spindle is weak, it is not negligible.

Recently developed automated systems can be used to measure the micronucleus frequency and gamma-H2AX (phosphorylation of histone is an indicator of DNA double-strand breaks) in the interphase nucleus. This is a promising method that allows analysis of a large number of subjects in cases of nuclear accident.[9] An additional advantage of the method is that only a small blood sample is required, enabling analyses to be carried out in small children.

Figure 13.3 Metaphase with chromosomes 1, 2 and 4 painted by FISH. Translocation of chromosome 1 is highlighted.

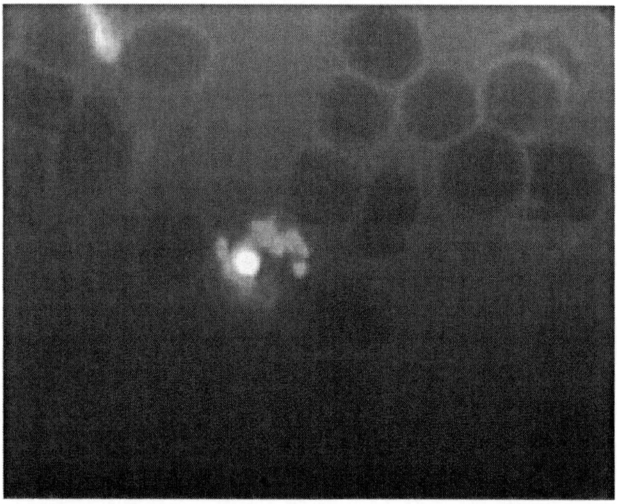

Figure 13.4 Reticulocyte with micronucleus.

Observation of micronuclei in peripheral reticulocytes is a relatively new method, and with the introduction of an automated system becomes a promising method in providing an estimation of genome damage within only a few hours of sampling and requiring only small samples of peripheral blood (Figure 13.4). The method is simple, does not require cell culture and has been proven to be accurate in estimating genome damage following exposure to ionizing radiation. As such this method is currently under investigation for human radiation biodosimetry.[10]

An evaluation of genome damage can also be performed in interphase cells where premature condensation of chromosomes (PCC) is induced. The advantage of this method is that cell culture is avoided and the dose can be determined immediately following exposure. This method can also be accompanied by FISH and could be applied after uniform and whole body exposure.[11]

During the last two decades, the Comet assay has been used as a biomarker of genome damage in biomonitoring following exposure to ionizing radiation. The disadvantage of this method is that it is exceptionally sensitive, possibly suggesting higher interindividual differences among the subjects within the control group than between the exposed and referent group. On the other hand, the method's sensitivity may be used in estimating individual radiosensitivity and repair capacity. The advantage of the method is that genome damage can be estimated within a few hours of sampling and it requires only a small blood sample.[12]

Minisatellite mutations induced by accidental exposure to ionizing radiation have been used in several studies as biomarkers of transgerenational effects. It was shown that minisatellite mutations may be used in detecting genome damage in germline cells and lymphocytes. However, although a significant increase in minisatellite mutations was detected, it was also shown that exceptional care should be taken in interpreting this type of biomarker because other genome–environment interactions affect the results of the biomarker.[13]

It is a well-known fact that the effects of ionizing radiation may be present and measured in remote cells, organs and tissues distant from the irradiated location. This mechanism, called the bystander effect, can be induced by both high and low LET radiations. It seems that multiple pathways are involved in this mechanism and different cell types respond differently to bystander signaling. Bystander cells express genetic and epigenetic changes, disturbed gene expression, activation of signal transduction pathways and delayed effects in their progeny.[14] The signal molecules that possibly play a role in intercellular communication are short RNAs, interleukins and the mitogen-activated protein kinase (MAPK) signaling cascade. In contrast to the bystander effect, ionizing radiation can also induce an adaptive response, which has been proven to be effective in reducing genome damage and bystander mutagenesis.

Despite the clear impact of bystander and adaptive mechanisms on the final biological effects of ionizing radiation, owing to their complex nature they have not as yet been included in the interpretation of results obtained by biomarkers.

Additionally, radiation may induce genome instability, which is a hallmark of cancer cells. Genome instability can be seen in persistent enhancement in the mutation rates and chromosomal aberrations caused in the progeny of the irradiated cells after many generations of replication. This phenomenon represents a serious problem in interpreting dose–response curves based on biomarkers because it causes the response profile to deviate from linearity.[15]

13.2.1 Protein and Amino Acid Biomarkers of Ionizing Radiation Exposure

Using high-resolution surface-enhanced laser desorption and mass spectrometry it is possible to determine the composition of proteins in serum, including the interleukin-6 precursor protein which changes following exposure to ionizing radiation in a dose-dependent manner. Twenty-three protein fragments/peptides can be determined in subjects exposed to ionizing radiation. This type of personalized proteomic analysis is a step forward towards high-throughput personalized biodosimetry.[16]

The difference between total body and organ irradiation can be measured by identifying different proteins (fetuin-B, tissue kallikrein, beta-glucuronidase chondroitin sulfate proteoglycan NG2, RNA binding protein 19, neuron navigator, Dapper homolog 3) in urine. Similarly, amino acids such as glycine and hydroxyproline reach 10 times their normal levels 1 week following ionizing radiation exposure. The advantage of these biomarkers for biomonitoring is their use in screening large populations, because urine analysis is a fast and non-invasive method.

13.2.2 Gene Expression after Exposure to Ionizing Radiation

During the last few years, significant efforts have been made in developing rapid and high-throughput radiation biodosimetry methods. Genes have been selected that respond to ionizing radiation and it is possible to apply about 50 of them as biomarkers.[17] Measuring mRNA is a very sensitive process and shows a linear dose–response relationship. The disadvantage of these biomarkers is that they can be measured for only a few days after exposure.

13.2.3 Tooth Enamel as a Biomarker of Exposure to Ionizing Radiation

Tooth enamel analysis is a method based on the measurement of free radicals produced and entrapped in the crystalline structure of the tooth after exposure to ionizing radiation. This elegant method detects this effect years after exposure. A linear dose relationship is available for this method. The method continues to be applied to teeth in subjects exposed during the bombing of Hiroshima and Nagasaki, providing reliable information on the type of exposure and the distance of the subject from the radiation source.[18]

Measurements of strontium-90 in teeth have been used over several decades in reconstructing internal and external doses. This biomarker is particularly interesting because it is non-invasive and can be used in measuring environmental and accidental exposure in children by utilizing baby teeth.[19] This same biomarker is particularly reliable in showing fluctuations of strontium-90 in the environment and indicating possible health risks even at low doses. Given that odontogenesis does cause tooth eruption, the method is very useful in estimating exposure during the post-pubertal period.

13.3 Biomonitoring of Tranplacental Exposure and Exposure of Children to Ionizing Radiation

Children are usually exposed to natural sources of ionizing radiation in their living environment. However, during wars or nuclear accidents, children may be exposed to various types and doses of ionizing radiation.

The principle effects of radiation on the developing embryo and fetus are growth retardation, embryonic, neonatal, or fetal death, congenital malformations and functional impairment, such as mental retardation.

Indoor radon, a natural source of ionizing radiation, is the most widespread type of exposure. Indoor exposure to radon at an annual dose of 7–11 mSv has been reported to be associated with a significant increase in the frequency of chromosome aberrations and micronuclei in children.

The largest amount of data available on genome damage in children following exposure to ionizing radiation has been from accidental overexposure incidents such as the Chernobyl, Techa River and Guaiana nuclear accidents, as well as medical therapy that include sources of ionizing radiation.

After the bombing of Hiroshima and Nagasaki, thousands of children were exposed to heavy radiation. Additionally, a large number of children were exposed *in utero* or born of parents exposed to heavy radiation. However, because cytogenetic methods were not available at the time, we have no information on the extent of genome damage. Studies measuring the exposed population began with the introduction of FISH. Conclusions cannot be made as to whether genome damage was more severe in children than in adults. Although these studies could not be performed in Japan, it has been shown in other cases that exposure during childhood causes greater permanent genome damage than in persons exposed during adulthood.

The largest body of data on genome damage in children exposed to ionizing radiation was collected after the nuclear plant accident in Chernobyl in 1986. Thousands of children were exposed to a large spectrum of doses (131I, 137Cs, 134Cs, 90Sr, a wide spectrum of short-lived isotopes). The studies were conducted on evacuated children and those born of evacuated mothers. National and international studies have confirmed that children are more radiosensitive. Given that the majority of the scientific papers on this topic were published in the Russian language, data transfer from knowledge gained progress slowly. A review of most of the relevant papers written in the English, Russian and Ukrainian languages on children exposed to ionizing radiation was published by Fucic *et al.* in 2008.[20] The conclusions of this paper are the following: (a) cytogentical biomarkers were applied in a majority of the studies; (b) children showed a higher radiosensitivity than adults; (c) genome instability and bystander effects were reported; (d) the presence of multiaberrant and rogue cells suggest internal contamination; (e) children exposed *in utero* showed increase genome damage; (f) genome damage after intrauterine exposure was observed mostly as chromosome inversions and deletions.

During the period 1948–1967, in the Southern Urals of the Soviet Union, industrial nuclear accidents took place in a facility known as the "Mayak

Complex". During this period, radioisotopes were released into the Techa River. Increased genome damage was observed in children of men who underwent exposure. There was a 1.7-fold increase in the minisatellite mutation rate in the germline of exposed fathers compared to referents from the rural area.

In the period 1949–1963 when nuclear tests took place at Semipalatinsk (Kazakhstan, in the former Soviet Union just south of the Altai region of Siberia), children were exposed to short-lived radionuclides present in food, water and air. Even 20 years following the cessation of nuclear testing, unstable chromosome aberrations such as rings and dicentrics were still detected in newborn infants, confirming the presence of genome instability in these subjects.

As a result of a Brazilian radiological accident in Goiania, an increased frequency of micronuclei was detected in children accidentally exposed to caesium-137.

Tooth enamel analysis has been used as a biomarker on several occasions. Significant interindividual differences were identified but the method was often successfully applied in practice to exposed children.

Intrauterine exposure of the fetus may occur if a mother is occupationally exposed to radioisotopes. Although pregnant women are usually isolated from hazardous jobs that may have a deleterious effect on the normal development of a fetus, some mothers are unaware of their first-trimester pregnancy and end up working during this sensitive period of fetal development in such environments. In hospitals where women are occupationally exposed to radioisotopes (iodine, thallium, technetium, thorium), the incidence of miscarriage increases significantly when compared with women exposed occupationally to X-rays, in which the miscarriage rate is the same as in the general population. Nonetheless, cytogenetic biomarkers show no significant difference in genome damage in women exposed to X-rays and radioisotopes. This provides evidence that in certain exposure settings, the induced health risk can be underestimated and other biomarkers should be applied or introduced.[21]

13.4 Biomonitoring after Occupational Exposure to Ionizing Radiation

A follow-up investigation using biomarkers in populations occupationally exposed to ionizing radiation is rarely incorporated in national legislations because it is assumed that physical dosimeters give reliable information on levels of exposure and health risk. The major health risk biomonitored in occupational exposure to ionizing radiation is cancer. Although it is true that groups of professionals such as physicians, nuclear plant workers and the police are usually exposed within recommended doses, still there are some occupations that require application of radioisotopes or mixed sources of ionizing radiation (industrial radiography, uranium mines, nuclear medicine departments, research units, the nuclear weapon industry, the radioisotope industry), where possible internal contamination needs to be additionally biomonitored. These subpopulations are shown to have increased genome damage despite the fact that physical measurements of ionizing radiation were within recommended dose levels. The reason for the detection of genome damage could be in fact that personal dosimeters only measure X- and gamma radiation, and these workplaces frequently reach permissible annual dose limits of 50 mSv (the average dose over a 5-year period should not exceed 20 mSv per year). Additionally, in some cases, subjects at workplaces are exposed to ionizing radiation, ultrasound and chemical agents for which personal dosimeters do not exist, and biomonitoring represents the only option in assessing the health risk.

Additionally, radiosensitivity in some persons may lead to more severe genome damage after exposure to ionizing radiation. It has been shown recently that increased genome damage following occupational exposure to ionizing radiation is detected in subjects with polymorphisms of the DNA repair gene XRCC3(241).[22] Before starting work in places exposed to ionizing radiation (with the exception of astronauts), testing of DNA repair capacity is usually not performed. These current results show the need for a professional consultation that includes the use of these and similar biomarkers to estimate an individual's radiosensitivity.

It has also been shown that the accumulated dose and the duration of employment are significantly correlated with genome damage. The application of a chromosome aberration assay for long-term low-dose exposure may underestimate the cumulative genome burden if FISH and translocation frequency analysis are not applied.

The importance of biomarkers in reporting a possible health risk prior to the appearance of clinical symptoms is evident when considering the case of healthy former nuclear-weapons workers who were exposed to plutonium many years ago and in whom increased frequencies of *intra*chromosomal rearrangements correlate with the plutonium dose in bone marrow. As *intra*chromosomal rearrangements are rare in a control population, a quantification of chromosomal damage in subjects exposed many years earlier could give new insights into the mechanisms and risks of cytogenetic damage.[23]

Aircrews are occupationally exposed to ionizing radiation, principally from galactic cosmic radiation. A significant increase of dicentric chromosomes and micronucleus frequency in pilots of the Concorde group has been detected, which seems to be a consequence of exposure to high LET cosmic radiation. Similarly, in astronauts exposed to low and high LET ionizing radiation over a period of 2 weeks to several months, increased levels of genome damage were detected, including translocations.

In summary, the application of biomarkers in estimating the health risk due to occupational exposure to ionizing radiation contributes significantly to better radiation protection policies and will in the future be the leading method in making particular decisions on the duration of employment in radiation zones.

13.5 Biomonitoring after Environmental and Accidental Exposure to Ionizing Radiation

In nature, organisms are exposed to terrestrial and cosmic forms of radiation. The major isotopes of concern for terrestrial radiation are uranium and its decay products such as thorium, radium, and radon. Low levels of uranium, thorium, and their associated decay products, potassium-40, carbon-14, and lead-210, are found everywhere. Some of these materials are ingested with food and water, while others, such as radon, are inhaled. The dose from terrestrial sources varies in different parts of the world.

Outside the Earth's atmosphere, galactic cosmic radiation consists mostly of fast-moving protons and alpha particles. After entering the atmosphere, these particles collide with the nuclei of nitrogen, oxygen, and other air atoms, generating additional ionizing radiation particles. The particles that enter the atmosphere and those generated are collectively referred to as galactic cosmic radiation. Doses from cosmic and terrestrial sources vary significantly.

During the last 100 years, humanmade radioactive sources, surface mines of radioactive radioisotopes, nuclear weapons testing and wars have significantly increased and changed the radioactive environment to which organisms on Earth have adapted during evolution.

Naturally occurring indoor radon-222 is associated with an increased risk of lung cancer. The accumulation of radon in homes, therefore, has led to a large program determining the effects of the densely ionizing alpha particles that are produced when radon decays. Increased levels of radon in indoor air (>200 Bq/m^3) were found to be linked to an increased level of DNA damage in peripheral lymphocytes.

The general population is exposed to ionizing radiation during diagnostic procedures such as X-ray imaging, computer tomography (CT), scintigraphy, *etc.* It has been shown that certain procedures such as cardiovascular interventions can induce increased the frequency of micronuclei in patients.

There are a large number of people exposed to ionizing radiation as a result of nuclear accidents in industry, and in nuclear test sites and wars. Although the consequences of nuclear accidents are also reflected by an increase in cardiovascular diseases, especially in the young population, because vascular endothelium has a very low repair capacity, biomonitoring of exposed populations is focused basically on identifying increased cancer risk.

Cytogenetic studies have shown increased genome damage in populations from the Chernobyl accident and other nuclear accidents in industry, or those exposed to isotopes handled accidentally by the general public or after exposure to radioactive cobalt incorporated by mistake into building structures. FISH has been shown to be the most informative method with regard to retrospective biodosimetry.

Biomonitoring of populations and soldiers exposed to depleted uranium during war activities are usually not performed, or are conducted on very small groups, but they indicate increased levels of chromosome aberrations.

13.6 Systems Biology as a Tool in Interpreting Present and Future Biomarkers

Health risk estimation based on an analysis of a single biomarker provides limited information on pathways and does not offer a complex picture of the network and interconnectivty of cells and tissues affected by ionizing radiation. Biomarkers used for identifying disturbances of the extracellular matrix do not exist, although the impact of mechanisms such as the bystander effect and adaptive response, which involve the extracellular matrix, are significantly related to the health risk.

Recently, it has been shown that a multi-parameter biological approach to estimating the health risk on the basis of the cytogenetic biomarkers Flt3ligand (hematopoietic biomarker), citrulline (digestive tract biomarker), and oxysterols (biomarkers of lipid metabolism and the vascular system) is of great importance in assessing radiation damage.[24]

Furthermore, the impact of other pathways, such as those directed by estrogen, should be included in the interpretation of biological effects after exposure to ionizing radtiation. It has been shown that the estrogen levels have a significant influence on carcinogenicity following exposure to ionizing radiation, and that such information is critical in the interpretation of biomarker results and health risk assessment.

The use of measurements of DNA methylation disturbance (epigenetics) after exposure to ionizing radiation are not included in biomonitoring studies, although this is one of the mechanisms associated with the bystander effect. Radiation causes DNA methylation changes which are dose dependent, and gender and tissue specific. It has been shown that organs not exposed directly to ionizing radiation also reveal different DNA methylation changes.

Gender difference in the response to ionizing radaitaion is a matter of future research. It has been shown recently that ionizing radiation causes various gender-related changes in the expression of microRNA (miRNA) in exposed brain tissue, which can be relevant in estimating cancer risk because clinical practice has shown that therapeutic use of ionizing radiation in males suggests a higher risk of developing brain tumors. Expectations are that in near future the so called miRNAome will become a new biomarker as a result of its sensitivity to low doses, and its dose-dependent expression pattern.

The systems biology approach utilizes genome-wide gene expression technologies and bioinformatics, which are especially important in estimating the health risk associated with low doses of radiation. Application of systems biology requires the introduction of a standardized language, enabling a better understanding of the processes involved on a graphical basis. This manner of combining graphics and symbols analysis of biomarkers is now moving closer to mathematics, the universal language. This allows scientists from different backgrounds to understand a common issue, and provides them subsequently with conditions for the optimal use of their knowledge.

As an example of this new approach for future biomonitorong of the effects of exposure to ionizing radiation, we provide a graphical presentation of the effects of uranium exposure (Figure 13.5).

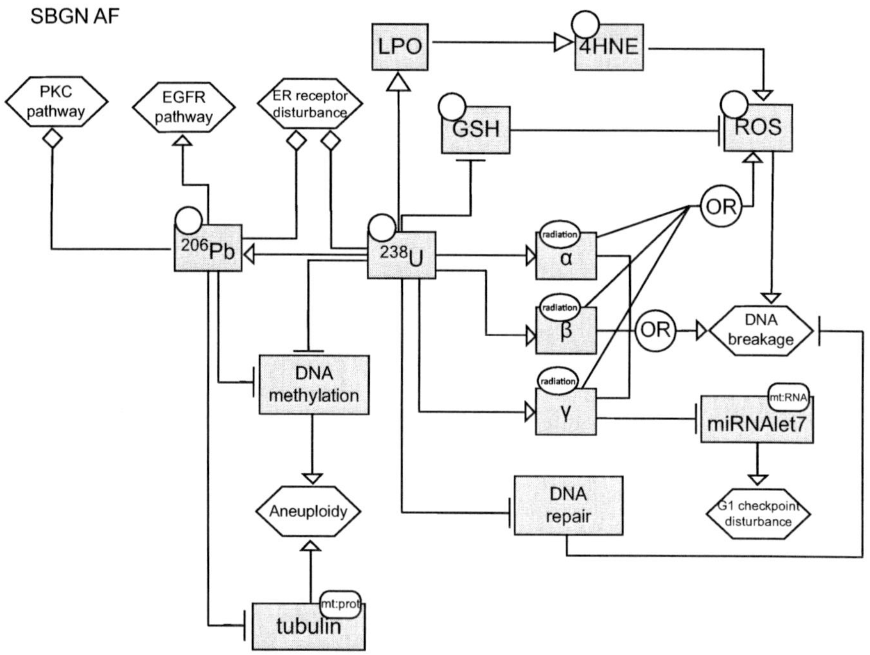

Figure 13.5 Graphical presentation of the biological effects after uranium exposure, using the Activity Flow language of Systems Biology Graphical Notation (SBGN-AF).

Uranium is a source of ionizing radiation and is also a heavy metal. Its xenoestrogenic activity is evident in estrogen disturbances, increases in lipid peroxidation and levels radical oxygen species, methylation disturbances, *etc.* The graphical presentation was prepared using the Activity Flow language of Systems Biology Graphical Notation (SBGN-AF).[25] The application of glyphs provides a picture of living system pathway complexity with regard to uranium exposure and the radioactive decay of lead, its daughter element. It also opens the possibility of revealing missing pathways that may hypothetically open up future research topics.

13.7 Conclusion

The final biological effects induced by ionizing radiation depend on the interplay between the DNA damage process and the signaling platform.

No matter where and which type of exposure to ionizing radiation is taking place in an environmental or occupational setting, the application of biomarkers has an advantage when compared to other physical measurements and calculations of health risk. However, each biomarker has its dose detection limits that should be followed in order to avoid false negative results (Table 13.2). Owing to interindividual differences in radiosensitivity, as well as specific age and gender related responses, biomarkers give the best insight into the

Table 13.2 The dose detection limit for major biomarkers applied after expsoure to ionizing radiation. FISH, fluorescent *in situ* hybridization; GPA, glycophorin A; HPRT, hypoxanthine-guanine phosphoribosyl transferase.

Biomarker	*Lowest dose detection limit (Gy)*
Chromosome aberration assay	0.5
Micronucleus assay	0.1
FISH	0.1
GPA variant analysis	1–2
HPRT mutant variant analysis	1–2
Gene expression changes	0.2

resulting biological effect. It is clearly important that several biomarkers are used in order to gain the most objective parameters in evaluating exposure effects or in preparing radiation protection legislation.

With regard to environmental exposure to ionizing radiation, there is a significant gap in biomonitoring of populations living close to nuclear plants. Although some studies have revealed an increased risk for childhood leukemia if the place of residence is closer than 5 km to a nuclear plant, up until now no full scale biomonotirng has been performed.

Terrorist attacks using low dose radiation sources represent a serious risk to the general population because such doses produce health deteriorations over a relatively long time period following exposure. Such exposure is identified only after the appearance of the first clinical symptoms. If such a terrorist attack were to be notified or announced by a terrorist organization, the mobilization of all available resources should, within a relatively short time, provide the possibility of performing biodosimetry on a large number of subjects. An effective protocol for this particular scenario does not exist.

The expectation is that the application of ionizing radiation for diagnostics will be reduced in time and, furthermore, there are future technological issues in avoiding the utilization of nuclear energy in electricity production.

The application of systems biology as a tool should in the future enable better interpretation of biomonitoring results and estimation of the health consequences of exposure to ionizing radiation on an individual basis.

References

1. S. Bonassi, H. Norppa, M. Ceppi, U. Strömberg, R. Vermeulen, A. Znaor, A. Cebulska-Wasilewska, E. Fabianova, A. Fucic, S. Gundy, I.L. Hansteen, L.E. Knudsen, J. Lazutka, P. Rossner, R. Sram and P. Boffetta, Chromosomal aberration frequency in lymphocytes predicts the risk of cancer: results from a pooled cohort study of 22 358 subjects in 11 countries, *Carcinogenesis*, 2008, **29**, 1178–1183.
2. J. N. Lucas, A. Awa and T. Straume, Rapid translocation frequency analysis in humans after exposure to ionizing radiation, *Int. J. Rad. Biol.*, 1992, **62**, 55–63.

3. U. Steinhaeser, H. Starke, A. Nietzel, J. Lindenau, P. Ullman, U. Claussen and T. Liehr, Suspension (S)-FISH a new technique for interphase nuclei, *J. Histochem. Cytochem.*, 2002, **50**, 1697–1698.
4. H. He, W. Deng, M. J. Cassel and J. N. Lucas, Fluorescence *in situ* hybridization of metaphase chromosomes in suspension, *Int. J. Radiat. Biol.*, 2001, **77**, 787–795.
5. R. C. Wilkins, H. Romm, T. C. Kao, A. A. Awa, M. A. Yoshida, G. K. Livingston, M. S. Jenkins, U. Oestericher, T. C. Pellmar and P. G. S. Prasanna, Interlaboratory comparison of the dicentric chromosome assay for radiation biodosimetry in mass causalty events, *Rad. Res.*, 2008, **169**, 551–560.
6. P. Bhatti, M. M. Doody, P. Rajaraman, B. H. Alexander, M. Yeager, A. Hutchinson, L. Burdette, G. Thomas, D. J. Hunter, S. L. Simon, R. M.Weinstock, M. Rosenstien, M. Stovall, D. L. Preston, M. S. Linet, R. N. Hoover, S. J. Chanock and A. J. Sigurdson, Novel breast cancer risk alleles and interaction with ionizing radiation among US radiological technologists, *Rad. Res.*, 2010, **173**, 214–224.
7. E. A. Ruiz-Narvaez, L. Rosenberg, C. N. Rotimi, L. A. Cupples, D. A. Boggs, A. Adeyemo, Y. C. Cozier, L. L. Adams-Campbell and J. R. Palmer, Genetic varaints of chromosome 5p12 are associated with risk of breast cancer in African American women: the Black Women's Health Study, *Breast Cancer Res.*, 2010, in press.
8. S. Bonassi, A. Znaor, M. Ceppi, C. Lando, W. P. Chang, N. Holland, M. Kirsch-Volders, E. Zeiger, S. Ban, R. Barale, M. P. Bigatti, C. Bolognesi, A. Cebulska-Wasilewska, E. Fabianova, A. Fucic, L. Hagmar, G. Joksic, A. Martelli, L. Migliore, E. Mirkova, M. R. Scarfi, A. Zijno, H. Norppa and M. Fenech, An increased micronucleus frequency in peripheral blood lymphocytes predicts the risk of cancer in humans, *Carcinogenesis*, 2007, **28**, 625–631.
9. G. Garty, Y. Chen, A. Salerno, H. Turner, J. Zhang, O. Lyulko, A. Bertucci, Y. Xu, H. Wang, N. Simaan, G. Randers-Pehrson, Y. Lawrence Yao, S. A. Amundson and D. J. Brenner, The RABIT: a rapid automated biodosimetry tool for radiological triage, *Health Phys.*, 2010, **98**, 209–217.
10. Y. Chen, Y. Tsai, I. Nowak, N.Wang, O. Hyrien, R. Wilkins, C. Ferrarotto, H. Sun and S. D. Dertinger, Validating high-throughput micronucleus analysis of peripheral reticulocytes for radiation biodosimetry: benchmark against dicentric and CBMN assays in a mouse model, *Health Phys.*, 2010, **98**, 218–227.
11. P. G. S. Prasanna, N. D. Escalada and W.F. Blakely, Induction of premature chromosome condensation by a phosphate inhibitor and a protein kinase in unstimulated human peripheral blood lymphocytes: a simple and rapid technique to study chromosome aberrations using specific whole-chromosome DNA hybridization probes for biological dosimetry, *Mutat. Res.*, 2000, **466**, 131–141. .
12. N. Touil, P. V. Aka, J. P. Buchet, H. Thierens and M. Kirsch-Volders, Assessment of genotoxic effects related to chronic low level exposure to ionizing radiation using biomarkers for DNA damage and repair, *Mutagenesis*, 2002, **17**, 223–232.

13. Y. E. Dubrova, Long-term genetic effects of radiation exposure, *Mutat. Res.*, 2003, **544**, 433–439.
14. O. Kovalchuk, F. J. Zemp, J. Filkowski, A. Altamirano, J. S. Dickey, G. Jenkins-Baker, S. A. Marino, D. J. Brenner, W. M. Bonner and O. A. Sedelnikova, MicroRNAome changes in bystander three-dimensional human tissue models suggest priming of apoptotic pathways, *Carcinogenesis*, 2010, in press.
15. W. F. Morgan and M. B. Sowa, Non-targeted effects of ionizing radiation: implications for risk assessment and the radiation dose response profile, *Health Phys.*, 2009, **97**, 426–432.
16. C. Menard, D. Johann, M. Lowenthal, T. Muanza, M. Sproull, S. Ross, J. Gulley, E. Petricoin, C. N. Coleman, G. Whiteley, L. Liotta and K. Camphausen, Discovering clinical biomarkers of ionizing radiation exposure with sreum proteoimc analysis, *Cancer Res.*, **66**(3), 1844–1850.
17. P. Sunirmal and S. A. Amundson, Development of gene expression signatures for practical radiation biodosimetry, *Int. J. Radiat. Oncol. Biol. Phys.*, 2008, **71**, 1236–1244.
18. W. Wallner, G. Ruhm, N. Rugel, A. Nakamura, T. Arazi, K. Faestermann, H. Knie, J. G. Maierf and G. Korschinekb, 41Ca in tooth enamel. Part I: A biological signature of neutron exposure in atomic bomb survivors, *Rad. Res.*, 2010, **74**, 137–145.
19. E. I. Tolstykh, E. A. Shishkina, M. O. Degteva, D. V. Ivanov, V. A. Shved, S. N. Bayankin, L. R. Anspaugh, B. A. Napier, A. Wieser and P. Jacob, Age dependencies of 90Sr incorporation in dental tissues: comparative analysis and interpretation of different kinds of measurements obtained for residents on the Techa river, *Health Phys.*, 2003, **85**(4), 409–419.
20. A. Fucic, G. Brunborg, R. Lasan, D. Jezek, L. E. Knudsen and D. F. Merlo, Genomic damage in children accidentally exposed to ionizing radiation: A review of the literature, *Mutat. Res.*, 2008, **658**, 111–123.
21. A. Fucic, D. F. Merlo, M. Ceppi and J. N. Lucas, Spontaneous abortions in female populations occupationally exposed to ionizing radiation, *Int. Arch. Occup. Environ. Health*, 2008, **81**, 873–879.
22. Y. H. Cho, Y. J. Kim, Y. S. An, H. D. Woo, S. Y. Choi, C. M. Kang and H. W. Chung, Micronucleus-centromere assay and DNA repair gene polymorphism in lymphocytes of industrial radiographers, *Mutat. Res.*, 2009, **680**(1-2), 17–24.
23. M. P. Hande, V. T. Azizova, R. C. Geard, L. E. Burak, C. R. Mitchell, V. F. Khokhryakov, E. K. Vasilenko and J. D. Brenner, Past exposure to densely ionizing radiation leaves a unique permanent signature in the genome, *Am. J. Human Genet.*, 2003, **72**(5), 1162–1170.
24. J. M. Bertho, L. Roy, M. Souidi, M. Benderitter, Y. Gueguen, J. J. Lataillade, M. Prat, T. Fagot, T. De Revel and P. Gourmelon, New bioloogical indicators to evaluate and monitor radiation-induced damage: an accident case report, *Rad. Res.*, 2008, **169**, 543–550.
25. N. Le Novère, M. Hucka, H. Mi, S. Moodie, F. Schreiber and A. Sorokin, The Systems Biology Graphical Notation, *Nat. Biotechnol.*, 2009, **27**(8), 735–741.

CHAPTER 14

Ethics and Data Protection in Human Biomarker Studies

B. DUMEZ,* K. VAN DAMME AND L. CASTELEYN

Katholieke Universiteit Leuven, Center for Human Genetics, Herestraat 49, 3000 Leuven, Belgium

14.1 Introduction

The use of biomarkers in environmental and health studies involves donation of tissues or fluids by volunteers and leads to sensitive ethical questions. To ensure the protection of the rights and dignity of study participants, the collection of human samples and data is subject to regulations and rules of different kinds, from deontological codes to data protection laws, from the local to the international level. Together with the decreased cost of analytical techniques and the increased possibilities for application of human biomarkers, the attention for, and interest in, the ethical and legal aspects has increased.[1–6]

Within a European Union (EU) context, probably the most important international references in this respect are the Data Protection Directive (95/46/EC),[7] the Oviedo Convention,[8] more in particular its Additional Protocol concerning biomedical research,[9] and Rec(2006)4 on research on biological material of human origin.[10]

The Data Protection Directive, which is subject to transposition in all EU Member States, concerns the processing of personal data and imposes, for example, the practice of informed consent, including the right to know one's own individual results, and notification of the research to the national supervisory authority. The Oviedo Convention and its Additional Protocol are documents open for signature and ratification by all Countries of the Council of Europe. They also emphasize the necessity of obtaining informed consent

Issues in Toxicology No. 10
Biomarkers and Human Biomonitoring, Volume 2: Selected Biomarkers of Current Interest
Edited by Lisbeth E. Knudsen and Domenico Franco Merlo

Published by the Royal Society of Chemistry, www.rsc.org

and require that a research project is submitted to an ethics committee for independent examination in every country where any research activity is to take place. Rec(2006)4 builds on the principles embodied in the Oviedo Convention and its Additional Protocol, but where the Additional Protocol covers interventions to remove biological materials for *specific* research projects and other interventions on human beings for *specific* biomedical research projects, or data collected for and resulting from these research projects, Rec(2006)4 focuses on the study of biological materials that have been *stored after originally being collected* in a diagnostic or therapeutic setting, during research projects with human subjects or during autopsy.

A key question is whether the rights and human dignity of study participants are adequately and equally protected throughout Europe and whether at the same time progress in environmental health-related studies is safeguarded. In other words, does the current ethical and legal framework protect what ultimately should be protected, and if so, to what extent may it need rethinking to achieve its dual goal: protecting both participants and scientific progress?

From a study of the general conditions in the current EU regulative and ethical framework, together with a critical analysis of research experiences in various EU countries,[5] it is shown that difficulties, ambiguities or even inconsistencies exist in the way ethical and juridical challenges are framed and being dealt with within and across countries. The diverse implementations of EU regulations or international guidelines in domestic law may in particular affect or hinder transnational research and bring about inequalities in the level of protection.

Even though many improvements regarding the legal and ethical challenges in human studies have already been implemented in the EU in recent years, these issues are not well known among the actors involved and efforts should be devoted to better education and dissemination of information and enhanced transparency targeting researchers but also the general public, the media and the policymakers.

In this chapter some of the challenges as well as steps to possible solutions are presented. More extensive information can be found in the ECNIS Volume on *Ethics and data protection in human biomarker studies*.[11]

14.2 Challenges

14.2.1 Inconsistencies and Fragmentation

Differences across countries in the transposition and implementation of EU regulations or guidelines in domestic law have consequences for transnational research.

Inconsistencies and difficulties in the current regulatory context are drawn as a result of the different interpretations which can be given – within and across countries – to key terms or concepts used in the privacy regulations such as "research", "anonymization", "controller", "substantial public interest", "disproportionate efforts", "compatibility of purposes",[12,13] and to the role attributed to ethics committees in the Oviedo Convention.

A reason for the sometimes not easily applicable concepts of the Data Protection Directive in the field of human biomarker practices in environmental health could be the fact that this Directive adopted a horizontal approach and applies to a very broad range of societal domains. Given that not every domain and its specificities can be covered, certain societal concerns can unwillingly become hindered such as the progress in research fields which contribute to people's health protection. It was of course impossible to foresee or to anticipate the implications of the Directive for each domain of society, and certainly for a particular domain such as human biomarker studies.

According to the Oviedo Convention, when data and/or samples are exchanged between countries in Europe for laboratory analysis, approval from a local ethics committee is required in some but not in all of the countries, often causing confusion for the researcher.

The creation of clarity and consistency in the interpretation of legal documents and in the implementation of an ethical and legal framework is not straightforward in the EU. Conflicting opinions exist regarding the need for more consistency. On the one hand, a call is launched for a more harmonized approach, including the use of standardized procedures, which increase comparability of results, but also tend to overlook local characteristics – while on the other hand the principle of subsidiarity is strongly recognized, thus promoting the development of procedures at a more local level, resulting in differences in approaches between Member States or research centers. Advancement of consistency is justified, but is not an easy endeavor. It should not necessarily entail a full or even partial standardization of methods or procedures. On the contrary, national traditions should be respected, though the level of protection of the participants should be equal throughout the Member States and the procedures transparent and unambiguous. If not, a shopping phenomenon may easily emerge: seeking to conduct research in the country with the least "resistance".

14.2.2 Individual and Public Concerns

A major issue for the specific field of human biomarker studies is the balancing of two different – and sometimes perceived as opposite – concerns: the need for the protection of the individual's privacy and the societal need for scientific progress. The tension between these two concerns should however not be overemphasized, because several studies highlight the willingness of study participants and populations to participate in studies with a view to promoting scientific knowledge in the interest of all. This is particularly so if participants are not merely considered as sample donators, but as full morally active participants in a research project. This consideration is of particular importance when addressing the challenges related to secondary use of data.

14.2.3 Secondary Use of Data

The secondary use of data or samples that were collected in previous research projects for other purposes sometimes remains very difficult or almost impossible. However, there are strong and recognized arguments for facilitation of such further use in the context of environmental health research where risks for improper use of data or samples are minimal, where public interest comes to the forefront and where adequate collective protection and control can be installed, so that breach of confidentiality or any use of data which would not be in line with the participant's moral stakes can be excluded.[14,15]

According to the Data Protection Directive, data may be re-used for a new (historical, statistical or scientific) purpose without renewed consent if "disproportionate efforts" are required to obtain such consent. The Additional Protocol to the Oviedo Convention specifies conditions for a research project designed for a specific purpose and does not foresee guidelines for the re-use of data and/or material. The Rec(2006)4 clarifies that if the original consent does not include the specified purpose for the re-use of data, all reasonable efforts should be made, both in terms of means and time, to re-contact the participant to request new consent. If re-contacting the participant is not possible despite reasonable efforts, Rec(2006)4 formulates specific conditions to comply with for the re-use of data. In practice, the decision whether a study objective is to be considered in line with the original purpose and whether the call for a new consent would require disproportionate efforts remains at first instance with the individual researcher. Such a decision-making process obviously adds to differences in practices within and across countries and may hinder consistency, and ultimately may delay and impede important research beneficial to all.

Between the two extremes of application of the current regulation, either following it in its strictest possible sense – making it sometimes impossible to process further valuable study material and thus hindering scientific progress – or adopting opportunistic strategies – perhaps ignoring the view of study participants – a more formal facilitation may increase the strength of EU regulations. However, caution should also be taken not to over-formalize research conditions because scientific progress might ultimately be hampered, without any real gain regarding the protection of the (potential) study participant.

When risks for the individual study participant are minimal and study results are expected to be profitable for the whole community, facilitation of secondary use can be implemented better, though based on firm conditions guaranteeing a full respect for human dignity through measures related to data confidentiality and regarding compliance with what can reasonably be assumed the motives of the participant for his or her original consent and his or her moral stake.[16–19]

14.2.4 Communication

Whilst human biomarker research may entail less or no risk for the study participant as compared to other applications that were at the origin of the establishment of the regulatory context of concern, and whilst it has its biggest merits in a common interest perspective, it requires an adapted approach to communication. More attention is needed – at the initiation of a study, during the study, as well as at the time of dissemination of results – for instance to the explanation of the meaning of the results at both individual level and collective level. At individual level this is important, for example to exploit the full potential of the tool in education and awareness raising, to deal with concerns and guilt feelings, to communicate findings which may have an impact on the health risks of the participant or indirectly for his or her relatives. At the collective level, it is important to express that the researchers are not sampling individual participants just for their own research interest, but with a view to contributing to the wellbeing of the population under study, thus encouraging their involvement also as a collective endeavor in eventual subsequent decision making on preventive action and policy measures.

14.3 Steps to Solutions

14.3.1 Information and Education

Even though many improvements regarding the legal and ethical challenges in human biomarker research have already been implemented in Europe in recent years, and possibilities have been created for allowing exemptions to the law for reasons of "substantial public interest" (art. 8 Data Protection Directive), thereby overruling the primacy of the individual in certain circumstances, the general need for a right balance between respect for the privacy of the individual and the need for scientific progress in the field of environmental health protection remains, and improvements in the field are needed. As stated above, in some cases there is a lack of clarity on the interpretation to be given to some key terms in the regulations. However, knowing that an all-embracing definition of terms such as "research", "disproportionate efforts", *etc.* does not exist, the fact that improved definitions could lead to a substantial facilitation of human biomarker studies may be seriously questioned. For example, the creation of a clear and all-encompassing definition of "substantial public interest" (mentioned in the Data Protection Directive) in the context of research is a difficult, if not an impossible, task. Nevertheless, it should also be mentioned that degrees of specification or definition may exist. For example, regarding the definition – or rather description – of "reasonable efforts", it can be said that whilst the Data Protection Directive and the Oviedo Convention and its Additional Protocol do not elaborate at all on what is to be understood under this term, Rec(2006)4 initiates a first attempt at a more specific description. Also, bodies such as the Article 29 Working Party, the independent advisory body on data protection and privacy, set up under Article 29 of the Data Protection Directive, try to examine questions covering the application of the national measures adopted under the EU directives in order to contribute to their uniform application. The Article 29 Working Party carries out this task by issuing recommendations, opinions and working documents. However these issues and instances are not well known among the actors involved, and efforts should be devoted to better education and dissemination of information and improved transparency targeting researchers but also the general public, the media and the policymakers. One way will obviously be the improvement of the communication processes related to human biomarker studies.

14.3.2 Communication and Raising Trust

The main endeavor is the reconciliation between respect for individual privacy and the progress of science. Whilst respect for data protection regulations could wrongfully be reduced to a simple administrative ritual, limiting oneself to a formal informed consent that may lack "authenticity", the key to making both privacy concerns and the concern for scientific progress not contradictory is building trust amongst (potential) participants and the general public in science, and thus in scientists, in their goals and in the context they are working in. Confidence can first of all be established by correct and respectful communication.

Respectful communication requires professionalism in many aspects, including the understanding by the recruiter how the confrontation of particular personalities may affect the (degree of authenticity of) the outcome of an informed consent procedure. Striving for an authentic informed consent, even knowing one can never reach 100% authenticity, is an ethical duty and highly valuable as such.[20]

It is important to realize that it is the way the study participant perceives the situation that will determine their attitude, their trust in science and thus their decision to participate or not. Correct communication presupposes both the willingness and ability or competence to communicate by those involved in executing a research project. Commitment and involvement of recruiters are preconditions for good communication, which should in turn be a prerequisite and an integral part of an ethically sound decision-making process. The vital role of communication is obvious: each communicative act of a researcher or any other person otherwise involved in human biomarker research may induce trust or distrust in a study participant. These experiences will be shared by the participant with others and eventually affect the perceptions of future potential participants. Involvement of communication experts in human biomarker studies is strongly advised. Researchers should organize and promote, and be actively involved in, democratic participation processes. These must be inseparable key principles in human biomarker research. Communication at a collective level with the population to which the study participants belong may be another key requirement in most biomarker studies. Informed consent of individual participants can never substitute for, and should be thus complemented by, this elementary step in a democratic participatory decision-making process.

14.3.3 The Role of Research Ethics Committees

Good communication in combination with confidentiality agreements alone will not be sufficient to make a research project ethically acceptable. Here is where the ethics committees should come in. While the rules from the Data Protection Directive guarantee notification and transparency, ethics committees should play a key role in assessing any problem which may arise regarding what might be called the reconciliation of privacy regulations and scientific needs. Ethics committees can fill in the regulatory gap regarding the assessment of the collective "societal" interest. They can provide clarification on what, for example, are "disproportionate efforts" or what should not be subject to the requirement of informed consent. Such decisions should indeed not be totally left to the researchers alone.

Ethics committees should have the authority and moral imperative to act for the good of study participants and their communities. They are in a crucial position to advocate thorough communication to diverse stakeholders, including the process of translation of research results into policy, thereby helping the concrete implementation by other social science expert groups. And, while focusing on different aspects, ethics committees should closely relate to and be in contact with national data protection authorities in order to benefit from each other's knowledge and competence. Ethics committees could substitute for individual informed consent in specific cases of secondary use of data and/or samples. This role could for example be made dependent upon the inclusion of a clause in the initial individual informed consent document in which the participant delegates the agreement for secondary use to an ethics committee. Questions like: "Do participants want to be re-contacted for re-use?" "Would an approval of an ethics committee suffice while the participant is notified through an information letter?" should then be included in the informed consent form.

14.3.4 A Research Ethics Committee at the European level

Ethics committees will weigh ethical principles against their reference regulations, which in most cases will be national regulations. Specifically for transnational research projects, to achieve a better consistency in the research conditions, the establishment of an ethics committee at the level of the "European research area" could be envisaged. Its main mission would be to give advice and/or decide on the interface between individual privacy regulations and concerns on the one hand and public health research needs on the other. Such a committee should have a permanent staff and ethical subcommittees or groups consisting of experts familiar with specific research domains. One expertise subgroup may for example involve human biomarker research in environmental health. The overall committee should survey consistency in the advising or decision-making processes of the subcommittees. The overall committee should report periodically to international institutions such as the European Parliament and the Council of Europe. Investment in such an effort could prevent loss of time, energy and research money. The analysis and evaluation of different cases in different situations will lead to a build-up of extensive knowledge and experience that may serve as an inspiring starting point for a well informed societal debate. This way, gradually, consistency in the handling of proposals and transparent decision making will be reached at the EU level, lifting the daily practices to a higher level. Formulated advice is deepened advice, possibly criticized or sharpened by national advice, and should be considered as a well thought out utterance of distilled jurisdiction and ethics. The European ethics committee would mainly be in place for transnational research, though it should be open for advice or opinion on national research projects as well. At the same time, the decision-making process should occur in a manner that provides access and participation to any person involved.

Through the implementation of such an EU research ethics committee, the inconsistencies in regulations across EU Member States regarding ethical acceptability of practices could be overcome. The very same initiative may also create opportunities for the elimination of the unnecessary bureaucratic burden for international research projects, and may aid considerably in avoiding irregular practices, while creating precedents that scientists can take into account when planning future research projects.

14.3.5 Participatory Approaches

A way of increasing the possibilities for the inclusion of arguments from the societal perspective lies in the current movement of participatory approaches.[6] Participatory processes make the public a partner in the communication process and are needed to combine technical expertise, rational decision making and public values and preferences. Mutual understanding among all actors involved and participation are necessary to create trust in order to solve problems that are both scientifically and socially complex. Striving for public participation in decision-making processes may promote research that is relevant to public health policy and the related translation of study results into action.[21,22]

14.3.6 Political Decision Making

Solutions must be framed through a political decision-making process. Politics should aim at the identification of societal problems and take the necessary measures to guarantee societal development, among which public health and scientific progress. Although elaboration on and the adoption of laws may seem the way forward, reality is very complex and no law or regulation will ever fully cover this complexity. Laws should therefore never be regarded as dogmas and should be evaluated continuously with respect to the basic societal and ethical concerns and principles that guided the legislator in the first place. Indeed, blind application of the law in cases of conflicting societal interest may lead to undesirable situations. Respect for legal provisions is necessary, but a critical attitude towards current legislation and the identification of drawbacks are equally important. This should lead to a societal debate and adaptation of the framework in place if appropriate.

14.4 Conclusion

Research ethics committees play a pivotal role in assessing to what extent decision-making processes fit with both individual and societal interest. Respect for human dignity and equality of moral status of all individuals, social justice, solidarity and democratic participation may thereby be appealing reflections of European values and useful complements to the four conventional bioethical principles (autonomy, beneficence, non-maleficence, and justice) which were initially conceived for a clinical context, but are widely referred to for evaluating policies, programs or activities that may entail risk to human health in general.

To achieve more consistency in the research conditions nationally and internationally an ethics committee at the level of the "European research area" could be envisaged, mainly focusing on transnational research. Moreover, the analysis and evaluation of different cases in different situations would build up extensive knowledge and experience that may serve as an inspiring starting point for a well informed societal debate. This way, gradually, more consistency in the handling of study proposals and increased transparency in decision making might be reached, lifting the daily practices and the protection of both individual and community interest to a higher level of meeting up with ethical concerns in transnational research.

An increasing demand exists for adapted communication strategies at all stages of a study, not only at the individual level, but also at the collective level, including the time of translation of results in preventive actions and policy making. The vital role of communication is obvious. Each communicative act may affect trust in the study at hand and in science in general. Expertise in the field of social sciences is therefore demanded.

Researchers have the duty to support the translation of research results into preventive actions at individual and at collective level whenever relevant. Participatory processes will facilitate the inclusion of arguments from the societal perspective and increase trust and mutual understanding among all parties involved, adding to the legitimacy of the final outcome and the public support for the decision-making policies.

Acknowledgement

This work was supported by the ECNIS Network of Excellence (Environmental Cancer Risk, Nutrition and Individual Susceptibility – Contract No. 513943) and NewGeneris IP (Contract No. 016320-2), operating within the European Union 6th Framework Program, Priority 5: "Food Quality and Safety".

References

1. J. Angerer, M. G. Bird, T. A. Burke, N. G. Doerrer, L. Needham and S. H. Robinson, Strategic biomonitoring initiatives: moving the science forward, *Toxicol. Sci.*, 2006, **93**(1), 3–10.

2. M. Arendt, Communicating human biomonitoring results to ensure policy coherence with public health recommendations: analysing breastmilk whilst protecting, promoting and supporting breastfeeding, *Environ. Health*, 2008, **7**(Suppl 1), S6.
3. O. Sepai, C. Collier, B. Van Tongelen and L. Casteleyn, Human biomonitoring data interpretation and ethics; obstacles or surmountable challenges? *Environ. Health*, 2008, **7**(Suppl 1), 13.
4. M. Harrison, Applying bioethical principles to human biomonitoring, *Environ. Health*, 2008, **7**(Suppl 1), S8. .
5. B. Dumez, K. Van Damme and L. Casteleyn, Research on the socio-ethical impact of biomarker use and the communication processes in ECNIS NoE and NewGeneris IP, *Int. J. Hyg. Environ. Health*, 2007, **210(**3-4), 263–265.
6. R. Morello-Frosch, J. Green Brody, P. Brown, R. Gasior Altman, R. A. Rudel and C. Pérez, Toxic ignorance and right-to-know in biomonitoring results communication: a survey of scientists and study participants, *Environ. Health*, 2009, **8**, 6.
7. Directive 95/46/EC of the European Parliament and of the Council of 24 October 1995 on the protection of individuals with regard to the processing of personal data and on the free movement of such data [cited 2010 Mar 1], available from: http://eur-lex.europa.eu/LexUriServ/LexUriServ.do?uri= CELEX:31995L0046:EN:HTML.
8. Convention for the Protection of Human Rights and Dignity of the Human Being with regard to the Application of Biology and Medicine: Convention on Human Rights and Biomedicine, 1997 Apr 4, Oviedo, Spain [cited 2010 Mar 1], available from: http://conventions.coe.int/Treaty/en/Treaties/Html/164.htm.
9. Convention for the Protection of Human Rights and Dignity of the Human Being with regard to the Application of Biology and Medicine: Convention on Human Rights and Biomedicine, 1997 Apr 4, Oviedo, Spain, available from http://conventions.coe.int/Treaty/en/Treaties/Html/195.htm.
10. Recommendation (2006)4 of the Committee of Ministers to EU Member States on research on biological material of human origin [cited 2010 Mar 1], available from https://wcd.coe.int/ ViewDoc.jsp?id=977859.
11. ECNIS, *Ethics and Data Protection in Human Biomarker Studies*, ed. L. Casteleyn, B. Dumez, A. Jamers and K. Van Damme, available from: http://www.ecnis.org/index.php?option = com_content&task = blogsection &id = 15&Itemid = 135.
12. D. Beyleveld, D. Townend and J. Wright, ed., *Research Ethics Committees, Data Protection and Medical Research in European Countries*, Ashgate Publishing Limited, Hampshire, UK, 2005.
13. D. Beyleveld and D. Townend, When is personal data rendered anonymous? Interpreting Recital 26 of Directive 96/45/EC, *World Congress on Medical Law*, 2002.
14. M. G. Hansson, Building on relationships of trust in biobank research, *J. Med. Ethics*, 2005, **31**, 415–418.

15. G. Helgesson, J. Dillner, J. Carlson, C. R. Bartram and M. G. Hansson, Ethical framework for previously collected biobank samples, *Nat. Biotechnol.*, 2007, **25**, 975.
16. WHO, *Guideline for Obtaining Informed Consent for the Procurement and Use of Human Tissues, Cells, and Fluids in Research*, World Health Organization, Geneva, 2003 [cited 2010 Mar 1], available from http://www.who.int/reproductivehealth/topics/ethics/human_tissue_use.pdf.
17. E. Salvaterra, L. Lecchi, S. Giovanelli, B. Butti, M. T. Bardella and P. A. Bertazzi, Banking together: a unified model of informed consent for biobanking, *EMBO Rep.*, 2008, **9**(4), 307.
18. C. K. Humphrey, C. A. Estabrooks, J. R. Norris, J. E. Smith and K. L. Hesketh, Archivist on board: contributions to the research team, Forum Qualitative Sozialforschung/Forum: *Qualit. Soc. Res.*, 2000, **1**(3), Art. 5 [cited 2010 Mar 1], available from: http://www.qualitative-research.net/index.php/fqs/article/viewArticle/1022/2203.
19. M. Law, Reduce, reuse, recycle: issues in the secondary use of research data, *IASSIST Quart.* 2005, **29**, 5–10.
20. A. Jamers, B. Dumez, K. Van Damme and L. Casteleyn, Video on 'Informed consent in environmental health research – A didactic tool for the improvement of the authenticity of informed consent', Katholieke Universiteit Leuven, Centre for Human Genetics, 2010.
21. F. Fisher, *Citizens, Experts, and the Environment: the Politics of Local Knowledge*, Duke University Press, Durham/London, 2000.
22. T. Webler, H. Karstenholz and O. Renn, Public participation in impact assessment: a social learning perspective, *Environ. Impact Assess. Rev.*, 1995, **15**, 443–463.

Subject Index

Locators in Roman typeface refer to volume 1
Locators in *italic* typeface refer to volume 2
Locators in **bold** refer to figures/diagrams
Locators for main headings which also have subheadings refer to general aspects of the topic